T0184292

Lecture Notes in Computer Science   11113

Commenced Publication in 1973
Founding and Former Series Editors:
Gerhard Goos, Juris Hartmanis, and Jan van Leeuwen

Editorial Board

David Hutchison
  Lancaster University, Lancaster, UK
Takeo Kanade
  Carnegie Mellon University, Pittsburgh, PA, USA
Josef Kittler
  University of Surrey, Guildford, UK
Jon M. Kleinberg
  Cornell University, Ithaca, NY, USA
Friedemann Mattern
  ETH Zurich, Zurich, Switzerland
John C. Mitchell
  Stanford University, Stanford, CA, USA
Moni Naor
  Weizmann Institute of Science, Rehovot, Israel
C. Pandu Rangan
  Indian Institute of Technology Madras, Chennai, India
Bernhard Steffen
  TU Dortmund University, Dortmund, Germany
Demetri Terzopoulos
  University of California, Los Angeles, CA, USA
Doug Tygar
  University of California, Berkeley, CA, USA

More information about this series at http://www.springer.com/series/7411

Massimo Coppola · Emanuele Carlini
Daniele D'Agostino · Jörn Altmann
José Ángel Bañares (Eds.)

Economics of Grids, Clouds, Systems, and Services

15th International Conference, GECON 2018
Pisa, Italy, September 18–20, 2018
Proceedings

 Springer

Editors
Massimo Coppola (iD)
National Research Council of Italy
Pisa, Italy

Emanuele Carlini (iD)
National Research Council of Italy
Pisa, Italy

Daniele D'Agostino (iD)
National Research Council of Italy
Genoa, Italy

Jörn Altmann (iD)
Seoul National University
Seoul, Korea (Republic of)

José Ángel Bañares (iD)
University of Zaragoza
Zaragoza, Spain

ISSN 0302-9743   ISSN 1611-3349 (electronic)
Lecture Notes in Computer Science
ISBN 978-3-030-13341-2   ISBN 978-3-030-13342-9 (eBook)
https://doi.org/10.1007/978-3-030-13342-9

Library of Congress Control Number: 2019931954

LNCS Sublibrary: SL5 – Computer Communication Networks and Telecommunications

© Springer Nature Switzerland AG 2019
This work is subject to copyright. All rights are reserved by the Publisher, whether the whole or part of the material is concerned, specifically the rights of translation, reprinting, reuse of illustrations, recitation, broadcasting, reproduction on microfilms or in any other physical way, and transmission or information storage and retrieval, electronic adaptation, computer software, or by similar or dissimilar methodology now known or hereafter developed.
The use of general descriptive names, registered names, trademarks, service marks, etc. in this publication does not imply, even in the absence of a specific statement, that such names are exempt from the relevant protective laws and regulations and therefore free for general use.
The publisher, the authors and the editors are safe to assume that the advice and information in this book are believed to be true and accurate at the date of publication. Neither the publisher nor the authors or the editors give a warranty, express or implied, with respect to the material contained herein or for any errors or omissions that may have been made. The publisher remains neutral with regard to jurisdictional claims in published maps and institutional affiliations.

This Springer imprint is published by the registered company Springer Nature Switzerland AG
The registered company address is: Gewerbestrasse 11, 6330 Cham, Switzerland

Preface

This volume constitutes the proceedings of the 15th International Conference on the Economics of Grids, Clouds, Systems, and Services (GECON 2018). GECON is a long-held conference that annually brings together economics and computer science researchers, with the ultimate aim of building a strong multidisciplinary community in the increasingly important areas of future ICT systems and economics.

The strong connection between ICT and economics aspects in current research endeavors reflects the reality of today's world, where technological advancements cause social and economic shifts, and information technology is often provided or controlled by large industrial realities and pervasive social networks. Clouds, mobile and distributed systems are well-established game-changer innovations of this kind. Other emergent ICT paradigms such as microservices and artificial intelligence are obvious examples of this recurring trend. Therefore, a multidisciplinary approach is needed when tackling the complexity of the interdependencies between ICT and economy, as well as the links with other disciplines such as politics or sociology. From another viewpoint, the impact of ICT is becoming so significant compared with other disciplines that, to some extent, the economy itself is transforming into an information and knowledge economy.

In order to better focus on new research directions and their impact and relationship with the main conference, this year's GECON organized three special sessions on selected topics, namely: "IT Service Ecosystems Enabled Through Emerging Digital Technologies," "Machine Learning, Cognitive Systems, and Data Science for System Management," and "Blockchain Technologies and Economics."

GECON 2018 was held during September 18–20, 2018, hosted by the Institute of Information Science and Technologies "A. Faedo" (ISTI) and located on the premises of the research area of the National Research Council of Italy (CNR) in Pisa, Italy. Pisa is a small but very lively city in Tuscany with an ancient history. Most notably, Pisa was one of the Maritime Republics in the Middle Ages, and is known worldwide for the Piazza dei Miracoli and its iconic Leaning Tower. Pisa is also a longtime hub of research excellence and higher education, being home to one of the oldest universities in Italy, as well as of the Scuola Normale Superiore, of the Sant'Anna School of Advanced Studies, and of the largest research area of the CNR.

We received 40 regular submissions for this year's edition. Each submission was assessed by three to five reviewers of the international Program Committee. Of these submissions, 11 were selected as full papers, for an acceptance rate of 27%. Additionally, nine shorter work-in-progress papers were integrated in the volume. Like in previous GECON editions, a combination of full papers and work-in-progress ones fulfills the twofold aim of gathering solid, original work and capturing innovative results. Starting from the presentation schedule and onward, the conference enabled open and informed dialogue between presenters and the audience, encouraging discussions and debates after each talk and providing enough discussion time.

Keynotes

This year's GECON featured three keynotes on ICT topics that deeply affect the way we use clouds as well as their economics.

The first invited speaker, at the conference opening, was Prof. Antonio Cisternino, University of Pisa. Antonio Cisternino has been a researcher at the Department of Computer Science since 2006. Among the positions he served there, he was the director of the Centre for Calculus of the department and participated in the constitution of the interdepartmental centre for research and services IT Center, of which he has been vice director since 2013. His main research interests include programming languages, meta-programming and domain-specific languages. He contributed to the .NET platform dynamic compilation runtime and the F# programming language. At present, he has turned his attention to clouds and virtualization, Fog Computing, and the Internet of Things. Also very active in technology transfer, Dr. Cisternino oversees the IT Center's collaborations with numerous industrial entities such as Ferrari, Microsoft, Dell, Intel, Acer, AMD and HP.

Prof. Cisternino's keynote "How Does the New Memory Hierarchy Affect the Cloud Model" tackled the issues and changes that recent developments of ITC hardware are bringing to the use and management of computing resources, both from a general viewpoint and specifically to the field of Cloud Computing.

"Cloud Computing, which started as a business model, has become a reference model to organize and use IT resources so that users are shielded by the ever-growing complexity of the ICT infrastructures. As any other model, the cloud one is based on implicit and explicit assumptions about the technology that made it possible. In this talk I present recent technology changes that are affecting the very underlying assumptions that have contributed to define the cloud reference architecture and the way cloud is used. In particular, I focus on the revolution of storage media and the impact on the whole communication infrastructure."

Also on the first day of the conference, the second invited speaker was Alessia Bardi, a researcher at Networked Multimedia Information Systems (NeMIS) Laboratory of CNR-ISTI. After getting her PhD in Information Engineering in 2016, Dr. Bardi has been involved in several EU-funded projects addressing the realization and operation of aggregative data infrastructures for various research communities. She is currently participating in projects supporting Open Access and Open Science, in particular OpenAIRE and OpenUP. Her research interests include service-oriented architectures, data and metadata interoperability, and data infrastructures for e-science and scholarly communication.

Dr. Bardi's keynote speech was about "Open Science as-a-Service for Repositories and Research Communities." A long abstract of her talk is included as the first contribution in this volume.

"Open Science is a set of practices of science mandating for accessibility and availability for re-use and re-distribution of research activities and products. Implementing Open Science principles aims at enabling responsible, reproducible, and transparently assessable research. Beside the need for a behavioral change in interested stakeholders, as of today the scholarly communication ecosystem lacks tools and open research community practices. To fill this gap and support a smooth transition toward Open Science, the OpenAIRE initiative is offering novel services for research communities and content providers."

The third invited speaker was Prof. Antonio Brogi. Antonio Brogi has been a full professor at the Department of Computer Science, University of Pisa (Italy) since 2004. He was previously there as associate professor and assistant professor, and later served as Dean of the Bachelor's and Master's Degree Programs in Computer Science. He leads the Service-Oriented, Cloud and Fog Computing research group (SOCC). His research interests include service-oriented and cloud-based computing, coordination and adaptation of software elements, formal methods and design of programming languages. He has published over 150 research papers and is member of the editorial board of several journals in the field, as well as member of the steering committees of conference series like CIbSE, ESOCC, and FOCLASA. The keynote from Prof. Brogi, on the second day of the conference, was entitled "Microservices Everywhere?"

"In this talk we critically discuss the main characteristics of microservices and the potentially huge advantages offered by their adoption for managing enterprise applications. We also show how a simple formalization of the main properties of microservices can be exploited to drive the refactoring of existing applications."

Special Topic Sessions

IT Service Ecosystems Enabled Through Emerging Digital Technologies
Chair: Somayeh Khaghighi, Amsterdam University, The Netherlands.

New business ecosystems have become possible thanks to digital platforms such as cloud computing, Internet of Things, and wearable technologies. Digital businesses in these ecosystems continuously monetize, manage, and measure information as an asset for having a competitive advantage in the market. This special session focuses on new ways of integrating people, resources, processes, and technologies, impacting on the management of systems and resources, on the analysis and modeling of value creation, and on the sustainability of technologies.

Machine Learning, Cognitive Systems and Data Science for System Management
Chair: Aurilla Aurelie Arntzen, University of South-Eastern, Norway.

It is well recognized in this interconnected world that businesses, government, and people depend on reliable technical infrastructures for all aspects of daily operations such as retail distribution, public and private transportation, and even social interaction. There are many reasons why failures occur in digital infrastructures and information systems, including human errors and malicious behavior. Today, the emerging trend is to use techniques from data science, Machine Learning, and Artificial Intelligence in order to automate management tasks, thus increasing system reliability and management efficiency.

Blockchain Technologies and Economics
Chairs: Paolo Mori, CNR-IIT, Italy and Stefano Bistarelli, University of Perugia, Italy.

The blockchain technology is raising increasing expectations, with many promising applications being proposed in several fields that go far beyond the crypto-currency use case that popularized the technology. This special session focuses on theory and applications of blockchain systems and services, as well as their impact on the viability of new economic models and issues on essential properties of the affected systems, including the legal, privacy, and security aspects.

Acknowledgments

We would like to thank first all the authors who submitted their work for evaluation. The community is what makes the conference possible. We would also like to wholeheartedly thank the reviewers and all Program Committee members for completing their reviews on time and providing insightful and valuable feedback to the authors. We wish to thank the invited speakers, for bringing new viewpoints and inputs to the GECON community, as well as the special session chairs, for their additional work and contribution to the conference. Furthermore, we would like to thank Alfred Hofmann and the whole team at Springer for their support in publishing the proceedings of GECON 2018. The collaboration with Springer was, as in the past, efficient and effective. Finally, we wish to acknowledge the support of the HPC Laboratory at ISTI-CNR and thank the ISTI-CNR administration team. Last but not least, we wish to acknowledge Asti Incentives & Congressi Srl, whose team supported the conference organization and handled most of the logistics.

October 2018

Massimo Coppola
Emanuele Carlini
Daniele D'Agostino
Jörn Altmann
José Ángel Bañares

Organization

Conference General Chairs

Massimo Coppola   ISTI-CNR, Italy
Emanuele Carlini   ISTI-CNR, Italy

Conference Vice Chairs

Jörn Altmann   Seoul National University, South Korea
José Ángel Bañares   University of Zaragoza, Spain
Karim Djemame   University of Leeds, UK
Congduc Pham   University of Pau, France

Proceedings Chair

Daniele D'Agostino   IMATI-CNR, Italy

Local Chair

Patrizio Dazzi   ISTI-CNR, Italy

Steering Committee

Jörn Altmann   Seoul National University, South Korea
José Ángel Bañares   University of Zaragoza, Spain
Steven Miller   Singapore Management University, Singapore
Mara Nikolaidou   Harokopio University of Athens, Greece
Omer F. Rana   Cardiff University, UK
Gheorghe Cosmin Silaghi   Babes-Bolyai University, Romania
Konstantinos Tserpes   Harokopio University of Athens, Greece

Program Committee

Alvaro Arenas   IE University, Spain
Aurelie Arntzen   University of South-Eastern Norway, Norway
Ashraf Bany Mohammed   The University of Jordan, Jordan
Stefano Bistarelli   University of Perugia, Italy
Ivona Brandić   Vienna University of Technology, Austria
Rajkumar Buyya   The University of Melbourne, Australia
Georg Carle   Technical University of Munich, Germany
Jeremy Cohen   Imperial College London, UK
Costas Courcoubetis   SUTD, Singapore

Daniele D'Agostino	IMATI-CNR, Italy
Patrizio Dazzi	ISTI-CNR, Italy
Alex Delis	University of Athens, Greece
Patricio Domingues	ESTG - Leiria, Portugal
Giancarlo Fortino	University of Calabria, Italy
Felix Freitag	Universitat Politècnica de Catalunya, Spain
Marc Frincu	West University of Timisoara, Romania
Saurabh Garg	University of Tasmania, Australia
Daniel Grosu	Wayne State University, USA
Netsanet Haile	Erasmus University, The Netherlands
Chun-Hsi Huang	University of Connecticut, USA
Bahman Javadi	Western Sydney University, Australia
Odej Kao	TU Berlin, Germany
Stefan Kirn	University of Hohenheim, Germany
Tobias Knoch	Erasmus University, The Netherlands
Bastian Koller	HLRS - University of Stuttgart, Germany
Somayeh Koohborfardhaghighi	University of Amsterdam, The Netherlands
Harald Kornmayer	DHBW Mannheim, Germany
George Kousiouris	National Technical University of Athens, Greece
Dieter Kranzlmüller	Ludwig Maximilian University of Munich, Germany
Dimosthenis Kyriazis	National Technical University of Athens, Greece
Joerg Leukel	University of Hohenheim, Germany
Dan Ma	Singapore Management University, Singapore
Richard Ma	National University of Singapore, Singapore
Leonardo Maccari	University of Trento, Italy
Ivan Merelli	ITB-CNR, Italy
Roc Meseguer	Universitat Politècnica de Catalunya, Spain
Mircea Moca	Babes-Bolyai University, Romania
Maurizio Naldi	Università di Roma Tor Vergata, Italy
Leandro Navarro	Universitat Politècnica de Catalunya, Spain
Marco Netto	IBM, Italy
Dirk Neumann	Albert-Ludwigs-Universität Freiburg, Germany
Frank Pallas	TU Berlin, Germany
George Pallis	University of Cyprus, Cyprus
Rubem Pereira	Liverpool John Moores University, UK
Dana Petcu	West University of Timisoara, Romania
Ioan Petri	Cardiff University, UK
Radu Prodan	University of Innsbruck, Austria
Peter Reichl	University of Vienna, Austria
Ivan Rodero	Rutgers University, USA
Rizos Sakellariou	The University of Manchester, UK
Benjamin Satzger	Microsoft, USA
Lutz Schubert	University of Ulm, Germany
Arun Sen	Arizona State University, USA
Jun Shen	University of Wollongong, Australia

Gheorghe Cosmin Silaghi	Babes-Bolyai University, Romania
Mathias Slawik	Medisite GmbH, Germany
Aleksander A. Slominski	Thomas J. Watson Research Center, USA
Burkhard Stiller	University of Zurich, Switzerland
Stefan Tai	TU Berlin, Germany
Rafael Tolosana-Calasanz	University of Zaragoza, Spain
Bruno Tuffin	Inria, France
Iraklis Varlamis	Harokopio University of Athens, Greece
Dora Varvarigou	National Technical University of Athens, Greece
Luís Veiga	Universidade de Lisboa, Portugal
Claudiu Vinte	Bucharest University of Economic Studies, Romania
Stefan Wesner	University Ulm, Germany
Phillipp Wieder	Gesellschaft für wissenschaftliche Datenverarbeitung mbH Göttingen (GWDG), Germany
Ramin Yahyapour	GWDG - University of Göttingen, Germany
Dimitrios Zissis	University of the Aegean, Greece

Additional Reviewers

Arvid Siqveland	Fernandes Steven
Badis Madani	Francesco Santini
Bhuvan Unhelkar	J. J. Villalobos
Carlo Taticchi	Kiran Raja
Carlos Pfeiffer	Narges Mehran
Damiano Di Francesco Maesa	Roland Mathà
David Bermbach	Valentina Giansanti

Contents

Special Topic Session - IT Service Ecosystems Enabled Through Emerging Digital Technologies

Special Topic Session - Machine Learning, Cognitive Systems and Data Science for System Management

Special Topic Session - Blockchain Technologies and Economics

Invited Papers

Open Science as-a-Service for Research Communities and Content Providers

Alessia Bardi$^{(\boxtimes)}$ iD

Institute of Information Science and Technologies – CNR, Pisa, Italy
`alessia.bardi@isti.cnr.it`

Abstract. Open Science is a set of practices of science according to which research activities and the research products they generate should be openly available, under terms that enable their findability, accessibility, re-use and re-distribution. The main effects of the implementation of Open Science principles is to enable responsible, reproducible and transparently assessable research.

For an effective implementation of Open Science principles, a behavioral change in interested stakeholders and new tools for publishing in the scholarly communication ecosystem are required. Open Science publishing calls for the publishing of all types of research artefacts, beyond scientific literature. Today, the scholarly communication ecosystem lacks of tools and research community practices on Open Science publishing. To fill this gap and support a smooth transition towards Open Science, the OpenAIRE initiative is offering two novel services for research communities and content providers (e.g., institutional repositories, data repositories). The final goal is to support the cultural and technological shift towards the Open Science paradigm, from which all the different stakeholders in the research domain and of the society at large can benefit.

Keywords: Open Science · Scholarly communication ·
Scientific communication · OpenAIRE

1 Open Science Publishing

Open Science is an umbrella term that defines a set of practices of science according to which research activities and any kind of products (e.g. literature, data, methods, software, workflows) they generate should be made available as soon as possible under terms that enable their findability, accessibility, re-use, and re-distribution [1]. Open Science principles are advocated by all scientific and scholarly communication stakeholders [2], who intend Open Science as a means for accelerating research by enhancing transparency and collaboration, and fostering innovation and reproducibility. Scientists and organizations see Open Science as a way to speed up, improve quality, and more effectively reward research activities, while funders and ministries see it as a means to optimize cost of science and leverage innovation [3].

Open Science is still an emerging vision and a way of thinking that calls for behavioral changes in the interested stakeholders and for novel tools in the scholarly

© Springer Nature Switzerland AG 2019
M. Coppola et al. (Eds.): GECON 2018, LNCS 11113, pp. 3–6, 2019.
https://doi.org/10.1007/978-3-030-13342-9_1

communication ecosystem that support the implementation of the Open Science principles.

Open Science publishing means publishing/depositing all products of a research activity under terms that enable their re-use and redistribution. The scholarly communication ecosystem should allow research communities to share and re-use their scientific results by publishing all intermediary and final research artefacts, beyond scientific literature. Artefacts can be research data, software and research methods (e.g. protocols, algorithms, etc.), which should be deposited in repositories for scientific communication (e.g. institutional repositories, data archives, software repositories, CRIS systems), and should be published together with the semantic links between them. To complete the picture, such ecosystem should support publishing of packages of artefacts (e.g. research objects [4], enhanced publications [5]) to allow discovery, evaluation, and reproducibility of science (e.g. workflows or experiments with input datasets). In addition, products of science should be published as soon as possible, i.e. not only once a research activity is concluded, but also while it is ongoing. In other words, scientists should, as much as possible, make their methodologies, thinking and findings available to enable/maximize collaboration and reuse by the research community and the society at large. An effective implementation of Open Science publishing principles would:

- Enable reproducible research: let other users reproduce experiments.
- Enable transparent evaluation of research activities: evaluate findings based on the ability to repeat science, but also on the quality of the individual products of science, i.e. literature, research data, software, and experiments.
- Support re-use of research results: overcome the barriers that SMEs and researchers (especially from developing countries) are facing in accessing publicly funded research.

1.1 Barriers to Open Science Publishing

The existing scientific communication ecosystem lacks tools and practices for engaging research communities at adopting the aforementioned novel Open Science publishing principles. Scientific literature, possibly linked to the underlying research data deposited in a data repository or archive, is still typically seen as the sole, omni-comprehensive unit of scientific dissemination. Publishing other types of artefacts like methods, protocols, workflows and software is not generally a common practice, i.e. no repositories of reference, no persistent identifiers, no citation practices and, therefore, no scientific reward. Regarding the publishing of packages of artefacts, solutions exists in discipline-specific communities (ISA Research Object for systems biology, MINSEQE and MIAMI guidelines for functional genomics) but, in general, research packages are not considered as first-class citizens of the scholarly communication domain.

The scientific ecosystem is also not able to keep a complete and up-to-date record of relationships between research products. For example, publication, data, software repositories and publishers do not keep bi-lateral links between their products, and the links they keep are not in-sync with the updates of the products (e.g. links to new versions of the data, obsolete links).

2 Open Science Publishing as-as-Service

OpenAIRE (Open Access Infrastructure for Research in Europe) [6] is an initiative funded by the European Commission that fosters and enables the adoption of Open Science publishing principles. Its aim is to provide Research Infrastructures (RIs) with the services required to bridge the research life-cycle they support - where scientists produce research products - with the scholarly communication infrastructure - where scientists publish research products - in such a way science is openly accessible, reusable, reproducible, and transparently assessable. OpenAIRE is working closely with existing RIs to (i) provide services for compensating the lack of Open Science publishing solutions and (ii) provide the support required by RIs to upgrade existing solutions to meet Open Science publishing needs (e.g. technical guidelines, best practices, Open Access mandates). To this aim, OpenAIRE is juxtaposing to its "Catch-all repository" Zenodo [7], two new services implementing the concept of "Open Science as a Service" (OSaaS), i.e. services and tools implementing Open Science principles are provided on-demand, transparently with respect to the underlying technical infrastructure. Tools are accessible through either a thin client interface, such as a web browser, or an application program interface. Services are designed to be usable by different disciplines and providers, each with different practices and maturity levels, so as to favor a shift towards a uniform cross-community and cross-content provider scientific communication ecosystem.

The *Research Community Dashboard* will serve research communities and their RIs to publish research artefacts, packages, and links, and to monitor their research impact. Scientists and infrastructural services populate and access an information space of interlinked objects dedicated to them, through which they can share any kind of products in their community, maximize re-use and reproducibility of science, and outreach the scholarly communication at large.

The *Catch-All Broker Service* will engage and mobilize content providers (institutional repositories, data archives, etc.) and serve them with services that help them in keeping their collections of scholarly records up-to-date. The idea behind the service is to disseminate and advocate the principle that scholarly communication data sources are not a passive component of the scholarly communication ecosystem, but rather active and interactive part of it. They should not consider themselves as thematic silos of products, but rather as hubs of products semantically interlinked with any kinds of research products and, more broadly, up-to-date with the evolving research ecosystem.

In order to facilitate the cultural and technological shift towards common Open Science publishing practices, the technological efforts are complemented with networking activities that will strengthen the emerging Open Science social environment, aligning practices and mechanisms for open accessibility, transparent evaluation and reproducibility.

3 Conclusion

The effective implementation of Open Science calls for a scientific communication ecosystem capable of enabling the Open Science publishing principles of transparency and reproducibility. Such ecosystem should provide tools, policies, and trust needed by scientists for sharing and interlinking (for "discovery" and "transparent evaluation") and re-using (for "reproducibility") all research artefacts produced during a research activity, e.g. literature, research data, methods, software, workflows, protocols.

OpenAIRE fosters transparent evaluation of results and facilitates reproducibility of science for research communities by enabling a scientific communication ecosystem where artefacts, packages of artefacts, and links between them can be exchanged across communities and across content providers. To this aim, OpenAIRE introduces and implements the concept of Open Science as a Service on top of the existing OpenAIRE infrastructure, by delivering services in support of Open Science publishing.

The final goal is to support the cultural and technological shift towards the Open Science paradigm, from which all the different stakeholders of the research domain and of the society at large can benefit in several ways, such as:

- Funders can optimize costs of research activities;
- Researchers can be more effectively credited and rewarded for all the research products they produce, not only for the scientific literature;
- Researcher's activities are reproducible and trustable;
- The open accessibility to all products enables transparent evaluation and support responsible research: SMEs, citizens and researchers worldwide, also from developing countries, can easily access research results, which, in Open Science settings, are not locked behind a paywall.

References

1. FOSTER Open Science Definition. https://www.fosteropenscience.eu/foster-taxonomy/open-science-definition. Accessed 16 Sept 2018
2. European Commission: Validation of the results of the public consultation on Science 2.0: Science in Transition [report]. Brussels: European Commission, Directorate-General for Research and Innovation (2015). http://ec.europa.eu/research/consultations/science-2.0/science_2_0_final_report.pdf
3. European Commission's Directorate-General for Research & Innovation (RTD): Open Innovation, Open Science and Open to the World (2016). https://ec.europa.eu/digital-single-market/en/news/open-innovation-open-science-open-world-vision-europe
4. Bechhofer, S., et al.: Why linked data is not enough for scientists. Future Gener. Comput. Syst. **29**(2), 599–611 (2013). ISSN 0167-739X. https://doi.org/10.1016/j.future.2011.08.004
5. Bardi, A., Manghi, P.: Enhanced publications: data models and information systems. LIBER Q. **23**(4), 240–273 (2014). http://doi.org/10.18352/lq.8445
6. Manghi, P., et al.: OpenAIREPlus: the European scholarly communication data infrastructure. D-Lib Mag. **18**(9), 1 (2012). https://doi.org/10.1045/september2012-manghi
7. Zenodo. https://zenodo.org/. Accessed 16 Sept 2018

General Track Papers

Applying Auctions to Bank Holding Company Software Project Portfolio Selection

David S. Gerstl[(✉)]

Department of Computer Systems, School of Business,
Farmingdale State College, The State University of New York,
2350 Broadhollow Road, Farmingdale, NY 11735, USA
GERSTLD@farmingdale.edu

Abstract. Large banks in the United States are often organized as bank holding companies controlling one or more subsidiaries. Their inorganic growth has led to duplicated capacity. In pursuit of cost savings, many holding companies provide shared services, including software development. Demand for development resources generally exceeds supply, leading to the need for a centralized project selection process. The selection methodology chosen is sometimes based on a resource constraint problem, using projected financials provided by the subsidiaries. This solution architecture is susceptible to misaligned incentives. We propose the use of a combinatorial auction to assist in project selection and more closely align the incentives of executives to those of the bank. We discuss appropriate auction formats for the contest based on the economic characteristics of the problem and explore impediments to implementation.

Keywords: Project selection · Combinatorial auctions · Shared-services

1 Introduction

Large banks are often organized as bank-holding companies (**BHC**s hereafter), financial holding companies controlling one or more banking subsidiaries. A wave of mergers has produced a diminishing number of ever larger BHCs, growing far in excess of secular industry growth, controlling banks doing a variety of consumer, investor, and corporate facing business [6]. This inorganic growth has led to duplicated capacity. As one major impetus for acquisition is cost reduction through operational synergies, BHCs move to centralize and/or outsource software development capabilities. While the eventual goal is to entirely centralize software development, in practice software development resources remain both in the subsidiaries and BHC, partially because acquisitive BHCs are never in a state of quiescence, so already integrated and organically grown subsidiaries retain capacity to avoid being at a strategic disadvantage. A major operational challenge for each BHC is deciding how to deploy centralized capacity in light of

© Springer Nature Switzerland AG 2019
M. Coppola et al. (Eds.): GECON 2018, LNCS 11113, pp. 9–23, 2019.
https://doi.org/10.1007/978-3-030-13342-9_2

requests exceeding aggregate capacity. Centralized project selection is required to fully realize efficiency gains [28] and avoid an "inverse tragedy of the commons" [17] where shared resources are underutilized due to diffused responsibility.

This work stems from management consulting work performed by the author for a major BHC. The BHC had, over time, acquired a variety of banking and financial services subsidiaries and made efforts to integrate and standardize acquisitions to achieve strategic and operational synergies. As noted in [28], in cases where "processes are ... being transformed", as after an acquisition, shared services are more incentive compatible than outsourcing. The BHC took a similar view and created a shared services development organization, using subsidiary-provided revenue projections to compare the returns of proposed projects. This naïve evaluation method was ripe for abuse, incentivising activity prioritizing the interest of the subsidiary and project executives over that of the BHC, such as overestimating expected revenue. While the evaluation of projects was out of scope in the bank engagement, and therefore the question was not addressed to the client, the misalignment inherent in this decision architecture was evident.

We make a number of simplifying assumptions. We assume that there is a single allocation metric, the programmer-week, with which we measure the required programming project investment, and that programmer-weeks are fungible. While we could easily scale capacity to account for variations in individual productivity, we instead ignore this. We ignore the scheduling of individual programmers and initially assume that each programmer works in only a single technology (relaxed in Sect. 2.2). We allow **free disposal**, where projects are permitted to use less than their allocated resources. We do not allow them to use more than allocated. We disallow resource transfer reasoning that this would be equivalent to permitting subsidiaries to run their own, internal, allocation process (although this may allow a robust, efficient secondary market to emerge). We assume that the software shared service may need to be a cost center, and that monetary transfers are used as a signaling but not necessarily a funding mechanism. We ignore BHC-wide synergies, and assume subsidiaries submit combinations of synergistic projects as mutually-exclusive singletons. These assumptions are simplistic, but consistent with those the author has observed in the field.

Prior work on project management has often centered on extensions to the standard project model [26], assuming that incentives among parties are aligned even when discussing shared services [28] or on optimization-based approaches to selection and scheduling [9]. The application of auctions to project selection has been proposed previously [29], possibly because of the connections between optimization and auction clearing algorithms [19], but often without regard to why it is preferable to a simpler optimization-based scheme. Clearing for combinatorial auctions and related optimization problems are well studied and provides a wealth of results that can be applied to auctions (e.g., [16]). Other work has examined **Enterprise crowdsourcing**, crowdsourcing taking place entirely within the enterprise, but given its need for a "scalable *expert* network" [33] (emphasis added) and the very different specialties of the banking subsidiaries, it may not be appropriate for picking projects for the shared-service.

In Sect. 2 we outline a project evaluation method similar to that used at the BHC. We follow this with a discussion of auctions as a method of discovering the internal valuation of a project in Sect. 3, including a comparison of some auction methods as applied to project selection. We then provide a short discussion on bid representation and computational complexity in Sect. 4. We give a simplified example in Sect. 5. Finally, we close with a summary and discussion of impediments in Sect. 6. For the convenience of readers, we **bold** terms in their first substantial use, near their definition.

2 Optimizing a Project Portfolio and Incentives

In order to abide by the ethical and legal restrictions imposed by the consulting relationship, we have abstracted out any details of the bank's method, leaving only the observation that subsidiaries provide projections used in an **ILP** (integer linear program) to allocate resources. Anecdotal evidence, from contemporaneous conversations, indicates that this evaluation scheme was not uncommon. The details contained here are thus an idealized version constructed using a standard project process [26] and an optimization approach, consistent with the observation that substantially all large firms use discounted cash flow and related methods for investment decisions [22]. Each project submission submits information about the project's expected revenues and expenses from which the BHC can compute an **NPV** (net present value). This process usually takes place among operating subsidiaries, some with their own development capabilities, a few levels below that of the top BHC. A simple example is shown in Fig. 1.

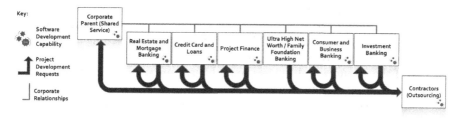

Fig. 1. Representative organizational structure and project development flows

2.1 An Optimization Based Approach

In designing a solution to allocate programming resources, we assume that each product can be created using a number of configurations of programmer-weeks. For example, a consumer facing product might be developed using different technologies, and might have specialized versions optimized for devices (e.g., a native iOS app). We assume that each proposer can determine their perceived value for each of these configurations and that the BHC will choose among these in the alternative. Given a number of subsidiaries proposing projects, the BHC can formulate an ILP to allocate development resources maximizing profitability [14]. With M different technologies denominated in some convenient unit (e.g., programmer-week). We denote I as the full set of available resources, C is an ordered integer M-tuple $\{C_1, \ldots, C_m\}$ denoting the capacity of each resource type in I, S a bundle of resources encoded as an ordered M-tuple $\{S_1, \ldots, S_m\}$ denoting the quantity of each resource type, J the set of projects, $v_j(S)$ denoting the value that project j places on bundle S (i.e., NPV for proposed bundles, 0 otherwise), and $x_j(S)$ indicates if j was awarded bundle $S \subseteq I$, then we solve:

$$\text{maximize: } \sum_{j \in J} \sum_{S \subseteq I} x_j(S) \cdot v_j(S) \qquad \text{(OBJ)}$$

$$\text{subject to: } \sum_{S \subseteq I} \sum_{j \in J} x_j(S) \cdot S_k \leq C_k \qquad \forall k, 1 \leq k \leq m \qquad \text{(1a)}$$

$$x_j(S) \in \{0, 1\} \qquad \forall S \subseteq I, \forall j \in J \qquad \text{(1b)}$$

(OBJ) encodes that we maximize the sum of the values of approved bundles, subject to: (1a), the sum total of any resource type contained in approved bundles is less than or equal to the capacity of that type; (1b), bundle allocation is atomic and therefore all $x_j(S)$ are binary variables. We model mutual exclusion restrictions (e.g., only one approved proposal per project) by adding a dummy unit capacity variable and using (1a). The related multi-unit winner determination problem [19] (**WDP**, hereafter), the problem of solving (OBJ) subject to (1), is known to be $\mathcal{NP}$-hard [18], but often amenable to heuristics [8,16,19]. One major advantage to this scheme is the ability of the subsidiary to control access to information, only disclosing resource signatures, value, and enough product detail for the shared service to be able to prevent project switching and preemptive resource bidding. We've eschewed imposing an academic formalism on the modeling of the resource problem as the problem representation is often dictated by an internal team without any exposure to formal methods in business process modeling. The basic model and process shown here are a natural enhancement of the existing data gathering and decision making process.

2.2   Programmers with More Than One Skill Set

As currently formulated, (1a) presumes that each programmer's capacity is assigned to a single technology, with the sums used to calculate the constants C_k. If all programmers are capable in the entire range of offered technologies, then (1a) can be configured as a single equation. More likely, (1a) must be modified to allow for the assignment of a programmer to multiple resource types without overutilization. For the purposes of exposition, assume that programmer 1 has 40 units to divide among technologies #1 and #2. Let C_1 and C_2 be the remaining capacity of technologies #1 and #2 respectively, and let U_1 and U_2 be the utilized capacity in a correct allocation. It is clear that if we add $U_1 - C_1 \leq 40$, $U_2 - C_2 \leq 40$, and $U_1 + U_2 - C_1 - C_2 \leq 40$, then (1a) can be

$$\sum_{S \subseteq I} \sum_{j \in J} x_j(S) \cdot S_k \leq U_k, \ \forall k, 1 \leq k \leq m \tag{1a-r}$$

This formulation does not work with more complex cases where programmers have overlapping, but not identical, skill sets. Instead, define variables P_j^i as the capacity assigned for programmer i for technology j, and add a constraint

$$P_1^1 + P_2^1 \leq 40 \tag{1c}$$

If programmer 2 has 25 units to allocate among technologies #2 and #3, then we add $P_2^2 + P_3^2 \leq 25$, and to limit the total assigned capacity to the correct U_k in (1a-r), we add $U_1 - C_1 - P_1^1 \leq 0$, $U_2 - C_2 - P_2^1 - P_2^2 \leq 0$, and $U_3 - C_3 - P_3^2 \leq 0$. See Sect. 5 for a partial example.

If more than one programmer shares the same set of skills, no additional variables or constraints are added but the constant in (1c) is adjusted to total capacity. Thus, for example, all the programmer who can cover every technology at the bank (the "all-arounders") would be captured with a single additional constraint of the form (1c). The number of additional variables and constraints is thus limited by the number of technologies supported, employee skill mix, and by corporate hiring and training practices, and is generally small in practice.

2.3   The Problem of Incentives

Market design is a subfield of economics that seek to efficiently bring together the parties to a transaction. One major reason why markets are redesigned is that the design is susceptible to **gaming**, strategic behavior on the part of participants at odds with the market designer's goals [27]. The BHC seeks to build the combination of software providing the greatest benefit to the bank subject to development resource constraints encoded in (1a). For this to work, the representations of value in (OBJ) as provided by the subsidiary (the **revealed valuation**) should reflect the internal value ascribed to the project by the proposer (the **true valuation**). Unfortunately, the existing design is subject to gaming.

Each proposal provides estimates to the BHC from which $v_i(S)$ is derived. The first is an estimate of programmer-weeks required. This estimate can be made using a variety of well-known methodologies, and the proposer is often in no favored position to know details significantly effecting the outcome. Additionally, development cost can often be observed within the year, making it unlikely that the proposer can expect any sustained benefit from a significant underestimate.

The estimate of projected revenue is more problematic. The future cash flow of a business without a track record is not much more than an educated guess derived from a plan, a market assessment, models, and assumptions. These estimates are within the area of expertise of the proposing subsidiary, but not the BHC's software development shared service. The timing of revenues makes this even more dangerous; the accuracy of revenue estimates may not be apparent for longer than the three-year average U.S. single company executive tenure [12].

3 Imposing Costs Early to Align Incentives

One solution to this temporal mismatch is to both allocate on the basis of the revealed valuation and immediately impose a commensurate cost on the successfully proposing subsidiary, giving their management a strong incentive to critically examine their submission. This method does not try to maximize revenue collected, but rather seeks to foster social efficiency, the optimal allocation of resources, which occurs where we maximize the sum of the surplus achieved by the BHC and the subsidiaries. Project i values resources S at $v_i(S)$. If it can purchase S for $p_i(S)$, then the **payoff** for purchasing S is $\pi_i(S) = v_i(S) - p_i(S)$. If we could determine $v_i(S)$, and charge proposer i that value, proposer i would be entirely indifferent as to whether to pursue the opportunity or not. Better would be to charge something less than $v_i(S)$ to give the proposer an incentive.

A well studied mechanism for allocating a scarce resource is the auction. One pricing rule yields a second-price auction (a.k.a. **Vickrey Auction**), in which the item(s) under auction is awarded to the highest bidder(s) at the price bid by the highest losing bidder [32]. The benefit of this pricing rule is that the **dominant strategy**, a strategy that does at least as well as any other, is to bid your true valuation [5]. A bidder who overbids risks paying above their true valuation and having a negative payoff. A bidder who underbids risks losing a potential positive payoff. A second price auction is thus **incentive compatible**, where participants get the best outcome by acting in accord with their true preference [5]. While we are unconcerned with maximizing the prices that the shared service achieves for its services as a means of subsidizing development, to the extent that the willingness to pay is a proxy for true valuation, an auction can effectively elucidate that value. In **common-value auctions**, those where the items are worth the same to every bidder, a well-known phenomena called the **winner's curse** exists where the bidder who most overestimates the value of the item wins, but overpays [31]. In cases where **individual private value**s predominate, such as here, the winner's curse is commensurately reduced [30].

3.1 Combinatorial Auctions and the VCG Auction

In situations where the multiple resources under auction are related, both the buyer (subsidiary) and seller (BHC) run significant risks with an item-by-item auction. The subsidiary is subject to **exposure risk** where that bidder wins "some—but not all—of a complementary collection of items ... The proposer is 'exposed' to a possible loss if his (sic) bids include synergistic gains that might not be realized" [10]. The BHC has an analogous problem. One solution is to auction items as **packages**, groups of items sold as a unit, using a multi-unit **combinatorial auction** [18] with salable item types with associated capacities. The Vickrey auction was first proposed for multiple homogeneous objects [32] and later extended to packages with heterogeneous objects piecemeal, becoming the **VCG auction**, after the initials of the proposers [5]. VCG uses the same formal constraints (OBJ) and (1). The difference between the optimization approach and the VCG is entirely in the adoption of a payment rule. With VCG, bidder i granted $S^i \subseteq I$ pays (under the Clarke pivot rule):

$$p_i(S^i) = \max \sum_{j \in J \setminus i} \sum_{S \subseteq I} x_j(S) \cdot v_j(S) - \max \sum_{j \in J \setminus i} \sum_{S \subseteq I \setminus S^i} x_j(S) \cdot v_j(S) \quad (2)$$

the externality that the winner imposes on others, that is the utility others would have if bidder i were absent from the auction but all items were unchanged, minus the utility they actually derived with both bidder i absent and only $(S - S^i)$ available [5]. First-price sealed bid auctions require the bidder to optimize the expected value of each bid, considering both the payoff and the probability of winning [21]. A **direct revelation mechanism** (e.g., VCG), where bidders bid their true valuations, enables bidding without this strategic computation.

In situations where the sale of multiple related items of value are considered, economists classify items as substitutes or complements, depending on how demand for [and therefore price of] one item changes as the other's price varies. **Complements** are pairs of items whose value in a combination may be higher than the value individually, and therefore one for which demand may fall when another item's price rises [10], the intuition being that if a project requires two different technologies then when the price of one goes up the project may be canceled, reducing demand for the second. **Substitutes** are items for which an increase in the price of one item does not reduce demand for the other [10]. These are often items that can both play the same role, so an increase in price for one may induce a project to shift to the other, increasing the second items price. In general, programmer-weeks can be complements to some bidders, substitutes for other bidders and may switch between complements and substitutes depending on price. The optimal auction format for substitutes and for complements is not the same. When a payment rule implies a payoff vector such that no subset of the bidders could approach the seller and defect (with them all doing at least as well, and some doing better), we say the result is in **the core**. With a VCG auction, (2) may not lead to a core allocation when goods are not substitutes [5].

Given the acquisitive manner in which the BHC brought many of its subsidiaries under its corporate umbrella, we expect some degree of collusion by projects from the same subsidiary. Collusive bidders act to reduce the prices paid by suppressing inter-bidder rivalry [21]. In the case of a single item Vickrey auction bidding rings have no effect unless the membership of the ring includes both of the bidders with the top two true valuations [21]. Even when the ring does include these members, the usual result is an adverse change in auction revenue, but no change in the assignment. As we are concerned with efficient allocation not with revenue maximization, this may be an acceptable result. Unfortunately, in a VCG auction, collusion can lead to an inefficient allocation [5]. The VCG auction, however, is unique in having certain positive properties, including a strong form of incentive compatibility that works regardless of whether other bidders are lying (interested readers are referred to [11] and the papers cited there). The loss of strong incentive compatibility is not fatal. There are weaker guarantees on bidder behavior that might be acceptable. For example, consider a guarantee that bidders use a semi-sincere strategy where they accurately give the relative value of packages, but may have an idiosyncratic discount they apply to each package. This may be the best we can do in the corporate context when looking at projects from across the enterprise. Large corporations overwhelmingly use hurdle-rates to handicap all projects before approval, but different subsidiaries often have different costs of capital and therefore different hurdle rates [22]. This implies that when comparing projects across the corporation we may need to settle for a semi-sincere strategy guarantee.

3.2   Ascending Proxy Auctions

Given potential bilateral relationships, where all parties have preference orders, the early 1960s saw the development of resource assignment algorithms based upon the concept of a stable marriage [13]. In this context, a **stable marriage** is a pairwise matching where, while each partner may have motive to defect and pair with other partners, the preference is not mutual so there is no opportunity for defection. One way to think about stable marriage is as a related concept to the core, indicating that no subgroup can profitably defect without some of them losing. In assigning graduating doctors to residencies and students to NYC high schools, a set of applicant-proposing deferred-acceptance algorithms based on stable marriages are currently used [1]. In these algorithms, both parties (e.g., students and schools) express a preference order and the algorithm makes a set of matches, moving in one direction from the perspective of resource holders (e.g., with the school's class monotonically improving) and the other direction from the perspective of supplicants (e.g., with the match moving down the student's preference list), permanently discarding matches from each list as it proceeds. The efficiency of this approach comes from the discard step, which ensures that the matching is limited to no more than the size of the preference lists. One failure of many matching algorithms is that they induce strategic behavior, for example when schools discriminate against students not ranking them first [1]. Here, the Department of Education acts as a mandatory proxy for the students, using the stated preferences to make the match without disclosing preferences, incentivising truthful disclosure and making it incentive compatible. This algorithm prevents defections by students and schools. An alternative design might optimize for finding an acceptable [non-core] assignment for students and disallow transfers outside the algorithm, but only once, as good students would learn to disclose only their most preferred choices, breaking incentive compatibility.

A more generalized version of this algorithm, called the ascending proxy auction [4], could be used to select projects. Ascending proxy auctions produce core outcomes when semi-sincere bidding is used [4]. The scalability of the high school matching algorithm is partially due to the identical resource signatures of every student. In the more general case, we may remove a project only to find sufficient resources for that project open up later. The result is that the ascending proxy auction can be exponential in its input. More problematic is the requirement (in common with VCG and stable marriage) that the bidder provide $v_i(S)$ for every acceptable package a-priori. There are an exponential number of potential packages and in the absence of pricing feedback, bidders focus on likely packages [3]. An iterative process allows "bidders to submit multiple bids during an auction and provides information feedback to support adaptive and focused elicitation" [24] ameliorating this concern.

3.3 Clock-Proxy Auctions

One solution to the related packaging, allocation, and pricing decisions that allows for pricing feedback is to separate these decisions into phases. The pre-selection of packages by the auctioneer is dangerous in that it can favor some bidders leading to inefficient allocations [5]. A number of related two-phase auction formats have been developed in the context of government spectrum auctions [10]. Here we adopt the clock-proxy auction [3]. In this approach, a clock auction phase is used to provide starting prices for each item type and create packages, and a second ascending proxy auction is used to determine the final configuration, prices, and allocation the packages. Each phase provides some advantages not present in a single phase auction [3].

Clock Round: During the clock phase, at each tick, the auctioneer announces prices on a unit bases, one price per type. Each bidder responds with a bid expressed as an item-demand vector. At the conclusion of each round, for items where supply exceeds demand, the price is fixed for the remainder of the clock phase. For those where demand exceeds supply, prices are increased. Bidders are only given the prices and aggregate demand for each item type, reducing the potential for gaming. The clock phase usually ends when excess demand has been extinguished. While it is possible to engineer a clock-proxy auction so that there is exposure risk for bidders [3], here all risk remains with the BHC. The clock phase is illustrated in Algorithm 1. We assume the presence of functions WDP solving the WDP and deltaPrices() finding the appropriate price change to an item with excess demand using one the suggested algorithms from Sect. 5.2.3 of [3]. At the conclusion of the clock phase, maxOBJ[] will be the clock-phase price vector maximizing auction revenue.

Proxy Round: The prices in maxPR[] become the minimum unit prices in the proxy round. Here, the auctioneer runs a mandatory proxy auction on packages, both those created in the clock phase as well as any new packages the bidders propose. At the conclusion of the proxy round a mutually compatible set of packages will have been won by bidders not subject to the exposure problem. Clock-proxy is relatively resistant to collusion [3].

In the clock round bidders who wish to minimize their own prices and maximize their chances of success (i.e., rational bidders) may be incented to both over and under-activity. Overactivity is exemplified by shill bidding to drive up the price of other projects' inputs. To dissuade shill bidding, every bid vector is recorded and does not expire, so the shill bidder may be held to those bids [3]. In underactivity, the bidder half-heartedly bids in the clock round to keep the prices of inputs low. Typically an activity rule, restricting bids based on prior

---

Data: prices[] : An array of initial prices, one per item type $1, \ldots, n$
Data: capacity[] : An array of capacity, one per item type $1, \ldots, n$

```
1  active[] := { 1, ..., 1 } ;              // all n item types are still in play
2  allBids := ∅ ;                           // Records all bids ever made
3  repeat
4  |   Publish the current prices prices[], receive bids ;   // See note in text
5  |   foreach bid B received this round do
6  |   |   allBids[] ∪ = {(prices[], B)} ;    // Store (price,bid) pair
7  |   |   demand[] += B ;                     // Compute Aggregate demand
8  |   if ∑ⁿᵢ₌₁(prices[i] · demand[i]) > ∑ⁿᵢ₌₁(maxPR[i] · maxDem[i]) then
9  |   |   maxPR[] := prices[], maxDem[] = demand[]; // save argmax of (OBJ)
10 |   foreach itemType i ∈ {1,...,n} do
11 |   |   if active[i] == 1 and (capacity[i] - demand[i]) > 0 then
12 |   |   |   prices[i] += deltaPrices(prices, capacity, bids, i);
13 |   |   else
14 |   |   |   active[i] := 0 ; // stop increase when demand matches capacity
15 until ∑ⁿᵢ₌₁ active[i] == 0;
```

Algorithm 1: Clock Phase adapted from [3]

---

activity, is used to prevent under activity, for example the **revealed preference activity rule** [3]. This rule restricts later bids to those bid earlier, unless the prices of *all* already bid packages have increased more than increase in the package now bid upon (possibly with some relaxation [25]). Intuitively, since each package S has a payoff for bidder i, if i is legitimately bidding on S' for the first time, it must be that *now* $v_i(S') - p(S') > v_i(S) - p(S)$ for all S previously bid, but since i didn't bid for S' earlier, it must also be that *at the time S was bid*, $v_i(S') - p(S') < v_i(S) - p(S)$. Since $v_i(S)$ and $v_i(S')$ don't change, it must be that $p(S)$ has increased more than $p(S')$ [3].

Pricing: We show an example and briefly discuss pricing in Sect. 5.

There are more complex choices possible in the design of our optimization framework by using an expanded set of metrics and data in the selection of a project portfolio. We've avoided adding additional parameters for two reasons. First, the parameters used here are all used already by the current process described in Sect. 2 so the details of their gathering are fixed and well understood. Adding additional parameters will implicate questions about data gathering procedures and would muddy the water when trying to evaluate a pilot. Second, many of the additional parameters that could be used are sourced from the subsidiaries and subject to the same misaligned incentives. We still allow the subsidiary to consider these parameters, but force them to summarize them in a single metric of expected value that aligns well with the metric that should be used by senior management. (i.e., "shareholder value").

4   Representation and Computational of the WDP

By design the clock phase is computationally simple. During the proxy phase, a solution to (OBJ) and (1) is required for each proxy round. While the proxy phase is ostensibly a series of rounds where bids are increased by ϵ, in practice the auction can be executed as a smaller number of discrete rounds. To express bids in a standard XOR bid specification language [23], each bid is encoded in the alternative, integrating optional features into the formula. So the bid, with utilization (u_k) and bid prices (p_k) integrated:

$$\emptyset \ [\ \oplus \ \overbrace{(p_a, \{u_1, \ldots, u_m\})}^{\text{Acceptable basic configuration}} \ \wedge \ \overbrace{(\emptyset \ [\vee \ (p_a^i, \{u_1^i, \ldots, u_m^i\}]^+)}^{\text{0 or more optional feature}} \ ]^+ \tag{3}$$

can be restated in an entirely disjunctive form as:

$$\emptyset \ [\ \oplus \ (p_a, \{u_1, \ldots, u_m\})][\oplus \ (p_a + p_a^i, \{u_1 + u_1^i, \ldots, u_m + u_m^i\})]^+ \ ]^+ \tag{4}$$

In theory this implies a possible exponential blowup in the number of bids as proposers try to convert instances of (3) into (4). In practice we expect bidders will act with restraint (possibly with an adverse impact on allocative efficiency). The general case for the solution to WDP is known to be $\mathcal{NP}$-hard [18,19] and not amenable to approximation over all forms of the problem. Nevertheless, with bid restrictions implied by the problem, there may be heuristics and approximation algorithms. Well known WDP_{xor} bid characteristics amenable to approximation and heuristics include submodular bids [19], small unit demand bids [7], and multi-minded bidders in cases with a small number of distinct goods [18]. Of these, only the final one applies, but that solution allows a bounded violation of the capacity constraint and will not work without significant swing capacity. Among the most promising methods for solving WDP is leveraging known results for branch-and-bound algorithms that have been shown to solve WDP [20] and related optimization problems in acceptable times.

5 An Illustrative Example

There is insufficient room for a full example with bidding rounds, but a short example will help elucidate some aspects of the auction process and pricing. Consider a [simplified] example with three technologies, four programmers and three proposed project as shown in Table 1. Omitting the S terms and (1b) for the sake of brevity, we add three variables in one instantiation of (1c): $P_{Java}^{1,2} + P_{Swift}^{1,2} + P_{SQL}^{1,2} \leq 60$. We omit the detailed bidding, except to look at how project 1A and 1B are both bid. Assuming that the price vector for the clock auction starts at $\{\$1, \$1, \$1\}$, initial aggregate demand will be at least $\{130, 0, 45\}$ (Project 1 prefers 1A). Eventually, when the price of 40 units of Java reaches at least \$12,030, $\pi_1(\{0, 30, 15\})$ exceeds $\pi_1(\{40, 0, 15\})$, and project 1 will shift its bids to 1B. This may be the first time 1B was considered, an example of the "adaptive and focused elicitation" mentioned in Sect. 3.3. Both sets of bids remain active indefinitely. In the end, the demand bid will imply an equation limiting Java: $x_{1a} \cdot 40 + x_2 \cdot 30 + x_3 \cdot 60 - 31 - P_{Java}^{1,2} \leq 0$, a similar SQL equation, and an [identity, discardable] Swift equation. To encode that projects 1A and 1B are mutually exclusive a dummy variable is added

Table 1. Simplified example: Three projects, four proposals

Programmer Capacity and Capability				
	Java	Swift	SQL	Weeks
Programmer 1	✓	✓	✓	30
Programmer 2	✓	✓	✓	30
Programmer 3	✓			31
Programmer 4		✓		32

Project requirements and value				
	Java	Swift	SQL	Value
Project 1A	40		15	$30k
Project 1B		30	15	$18k
Project 2	30		15	$20k
Project 3	60		15	$33k

and the equation reduces to $x_{1a} + x_{1b} \leq 1$. After the proxy round, (OBJ) is **Maximize** : $x_{1a} \cdot 30k + x_{1b} \cdot 18k + x_2 \cdot 20k + x_3 \cdot 33k$. The winning bids will be P_{1b} and P_3. Naïve application of (2) would result in Project 3 paying $(38k - 18k) = 20k$ and Project 1B paying $(33k - 33k) = 0$ for their allocation. The "zero revenue" problem is well known in VCG auctions [5]. It is instructive to consider why (2) yields zero when applied to 1b. The problem here is a supply-demand mismatch. There is a supply of Swift programmer-weeks with zero opportunity cost (since there is only one project with demand). Project 1b should pay nothing for this supply as it's perishable. Project 1b might pay for its demand for SQL, except there is no other project that can use that 15 programmer-weeks, so again there is zero opportunity cost. In a typical example within the bank, we'd expect to see a larger variety of projects, better matching of supply and demand (whether through programmer hiring, programmer training, or a realization by others that projects appropriately configured may be able to pay less that their value).

While combinatorial auctions generally use a second price model [2], the vast majority of actual auctions "repair" the pricing of (2) to bring prices paid within the core. While there is considerable more mechanism behind pricing rules, one key insight about core pricing is that "...any set of bidders pays at least as much as their opponents would pay to take their stuff away from them" [11]. If we consider the same example, but with 1a and 1b belonging to different projects (and conflicting on a single resource to keep $1x_{1a} + x_{1b} \leq 1$), any total revenue of less than $30k from 1b and 3 would not prevent 1a from successfully defecting, so the total revenue needs to be increased by $10k. [11] provides a model for how to distribute the excess charge among 1b and 3. This correction, however, may have an adverse effect on incentive compatibility [15].

6   Summary, Impediments, and Next Steps

We summarize the assignment methods discussed here in Table 2. The author is currently seeking a banking partner with which to pilot this method but, given the risk of data leakage of proprietary financials, is not optimistic about the possibility of publishing more than anecdotal results.

Two large impediment still exist. One big difficulty in applying this to shared service software development may be the secular migration to agile, which makes this type of command-and-control planning less certain, and may make this more

Table 2. Discussed allocation methods: (r) indicates only w.r.t. *reported* valuation

Allocation method	Allocative efficiency	Incentive compatible	Pricing feedback	Stable/Core seeking	Collusion resistant
Optimization	●(r)	○	○	N/A	○
VCG [5,32]	●	●	○	○	○
Stable marriage [1,13]	●	●	○	●	●
Ascending proxy [4]	●(r)	◑	○	●(r)	◑
Clock-proxy [3,11,15]	●(r)	◑	◑	●(r)	◑

applicable in the general corporate investment and shared service context (until those become agile). Another remaining difficulty in applying this methods to shared service prioritization is the presence of external substitutes in the form of outsourced development. While the BHC might prefer that subsidiaries utilized the shared service rather than outsourcing, the advantages of a shared service are limited. The shared service can develop software with reduced acquisition overhead and better knowledge and capabilities with the internal IT systems (to the extent they are integrated), but in a friction-free environment, the major advantage that the shared service retains is a lower price due to limited profit motive and/or subsidization. Thus, in considering $v_j(S)$ in (OBJ) we need consider not the true valuation of the project, but rather the maximum of that valuation and the external price. If this price is much less than $v_j(S)$, as with high NPV projects, the availability of this substitute means the auction may not separate the truly valuable from the merely profitable projects absent a mandate that makes the shared service a monopolist. One benefit of this parallel capacity is that the bank subsidiaries can be assured of best efforts by the shared service as there is an alternative market to which the subsidiaries can resort if they are not satisfied with their shared service performance. Another benefit is that the cost is somewhat constrained as the shared service must compete with the open market. Both of these may be absent if we mandate the use of the shared service. There are, however, two easy instances when there are no good substitutes and this method might apply without a heavy-handed mandate. First, some projects require a level of IT integration that the BHC might be unwilling to allow of an outside company. Second is the small project. While in a friction-free environment, even small projects can be contracted out, in the real world there is significant transaction costs to outsourcing smaller projects.

There are a number of issues we have not addressed. Much of the auction literature assumes that buyers are risk neutral, but there is good reason to believe that there is a systemic bias in the hiring of more risk tolerant employees in some divisions (e.g., investment bank) than in others (e.g., retail bank). We also haven't dealt directly with the accounting, compensation, and tax implications of moving profits from the subsidiary to the BHC, with risk concentration, with scheduling, with liquidity constraints, with product cannibalization, with corporate strategy, with regulatory projects, and with the question of whether the shared service is necessarily a cost-center (i.e., whether we can have a weak balanced budget constraint). These are all areas that can be explored in more depth in a different format.

References

1. Abdulkadiroğlu, A., Pathak, P.A., Roth, A.E.: The New York City high school match. Am. Econ. Rev. **95**(2), 364–367 (2005)
2. Ausubel, L.M., Baranov, O.V.: Market design and the evolution of the combinatorial clock auction. Am. Econ. Rev. **104**(5), 446–451 (2014)
3. Ausubel, L.M., Cramton, P.C., Milgrom, P.R.: The clock-proxy auction: a practical combinatorial auction design. In: Cramton et al. [10], Chap. 5, pp. 115–138
4. Ausubel, L.M., Milgrom, P.R.: Ascending auctions with package bidding. Front. Theor. Econ. **1**(1), 1–42 (2002)
5. Ausubel, L.M., Milgrom, P.R.: The lovely but lonely Vickrey auction. In: Cramton et al. [10], Chap. 1, pp. 17–40
6. Avraham, D., Selvaggi, P., Vickery, J.I.: A structural view of U.S. bank holding companies. Econ. Policy Rev. **18**(2), 65–81 (2012)
7. Bartal, Y., Gonen, R., Nisan, N.: Incentive compatible multi unit combinatorial auctions. In: Proceedings of the Ninth Conference TARK, pp. 72–87. ACM (2003)
8. Bettinelli, A., Cacchiani, V., Malaguti, E.: A branch-and-bound algorithm for the knapsack problem with conflict graph. INFORMS J. Comput. **29**(3), 457–473 (2017)
9. Chen, J., Askin, R.G.: Project selection, scheduling and resource allocation with time dependent returns. Eur. J. Oper. Res. **193**(1), 23–34 (2009)
10. Cramton, P., Shoham, Y., Steinberg, R. (eds.): Combinatorial Auctions. MIT Press, Boston (2006)
11. Day, R.W., Cramton, P.: Quadratic core-selecting payment rules for combinatorial auctions. Oper. Res. **60**(3), 588–603 (2012)
12. 2009 executive job market intelligence report. Technical report, ExecuNet (2009)
13. Gale, D., Shapley, L.S.: College admissions and the stability of marriage. Am. Math. Mon. **69**(1), 9–15 (1962)
14. Ghasemzadeh, F., Archer, N., Iyogun, P.: A zero-one model for project portfolio selection and scheduling. J. Oper. Res. Soc. **50**(7), 745–755 (1999)
15. Goeree, J.K., Lien, Y.: On the impossibility of core-selecting auctions. Theor. Econ. **11**(1), 41–52 (2016)
16. Kellerer, H., Pferschy, U., Pisinger, D.: Knapsack Problems. Springer, Heidelberg (2003). https://doi.org/10.1007/978-3-540-24777-7
17. Knoch, T.A., Baumgärtner, V., Grosveld, F.G., Egger, K.: Approaching the internalization challenge of grid technologies into e-Society by e-Human "Grid" ecology. In: Altmann, J., Rana, O.F. (eds.) GECON 2010. LNCS, vol. 6296, pp. 116–128. Springer, Heidelberg (2010). https://doi.org/10.1007/978-3-642-15681-6_9
18. Krysta, P., Telelis, O., Ventre, C.: Mechanisms for multi-unit combinatorial auctions with a few distinct goods. J. Artif. Intell. Res. **53**, 721–744 (2015)
19. Lehmann, D., Müller, R., Sandholm, T.: The winner determination problem. In: Cramton et al. [10], Chap. 12, pp. 297–318
20. Leyton-Brown, K., Shoham, Y., Tennenholtz, M.: An algorithm for multi-unit combinatorial auctions. In: AAAI/IAAI, pp. 56–61 (2000)

21. Marshall, R.C., Marx, L.M.: The Economics of Collusion: Cartels and Bidding Rings. MIT Press, Cambridge (2012)
22. Meier, I., Tarhan, V.: Corporate investment decision practices and the hurdle rate premium puzzle. In: 2006 Annual Management European Financial Management Association (2006)
23. Nisan, N.: Bidding languages. In: Cramton et al. [10], Chap. 9, pp. 215–231
24. Parkes, D.: Iterative combinatorial auctions. In: Cramton et al. [10], Chap. 2, pp. 41–78
25. Parkes, D.C., Rabin, M.O., Thorpe, C.: Cryptographic combinatorial clock-proxy auctions. In: Dingledine, R., Golle, P. (eds.) FC 2009. LNCS, vol. 5628, pp. 305–324. Springer, Heidelberg (2009). https://doi.org/10.1007/978-3-642-03549-4_19
26. Pinto, J.K.: Project Management: Achieving Competitive Advantage. Prentice Hall, Upper Saddle River (2010)
27. Roth, A.E.: Who Gets What—and Why: The New Economics of Matchmaking and Market Design. Houghton Mifflin Harcourt, Boston (2015)
28. Sako, M.: Outsourcing versus shared services. CACM **53**(7), 27–29 (2010)
29. Shou, Y.Y., Huang, Y.L.: Combinatorial auction algorithm for project portfolio selection and scheduling to maximize the net present value. J. Zhejiang Univ. Sci. C **11**(7), 562–574 (2010)
30. Tenev, A.P., Peeters, R.: Number of bidders and the winner's curse. Economic Discussion Papers No. 1802, University of Otago (2018). https://doi.org/10.1515/bejeap-2018-0025
31. Thaler, R.H.: Anomalies: the winner's curse. J. Econ. Perspect. **2**(1), 191–202 (1988)
32. Vickrey, W.: Counterspeculation, auctions, and competitive sealed tenders. J. Finance **16**(1), 8–37 (1961)
33. Vukovic, M., Laredo, J., Rajagopal, S.: Challenges and experiences in deploying enterprise crowdsourcing service. In: Benatallah, B., Casati, F., Kappel, G., Rossi, G. (eds.) ICWE 2010. LNCS, vol. 6189, pp. 460–467. Springer, Heidelberg (2010). https://doi.org/10.1007/978-3-642-13911-6_31

Preventing Collusion in Cloud Computing Auctions

Shunit Agmon[1(✉)], Orna Agmon Ben-Yehuda[1,2], and Assaf Schuster[1]

[1] Technion—Israel Institute of Technology, 3200003 Haifa, Israel
{shunita,ladypine,assaf}@cs.technion.ac.il
[2] Caesarea Rothschild Institute for Interdisciplinary Applications of Computer Science, University of Haifa, Haifa, Israel

Abstract. Cloud providers are moving towards auctioning cloud resources rather than renting them using fixed prices. Vickrey-Clarke-Groves (VCG) auctions are likely to be used for that purpose, since they maximize social welfare—the participants' aggregate valuation of the resources. However, VCG auctions are prone to collusion, where users try to increase their profits at the expense of auction efficiency. We propose a coalition formation mechanism for cloud users that helps both users and providers. Our mechanism allows the auction participants to collaborate profitably while also maintaining the auction's resource allocation efficiency. Our experiments show that when using our mechanism, participants' mean profit increases by up to 1.67x, without harming the provider's allocation efficiency.

Keywords: Cloud · Auctions · Collusion

1 Introduction

Cloud computing provides flexibility to clients by allowing them to pay per use for the rental of services and VMs. Renting reduces the waste of prepurchased but unutilized hardware [15]. Recently, cloud computing has been moving towards the more economical Resource-as-a-Service model (RaaS) [11,13]: instead of horizontal scaling (renting more VMs), RaaS clouds enable vertical scaling—renting more resources (such as CPU, RAM, and I/O resources) for a few seconds at a time, at sub-second granularity. For example, CloudSigma charges separately for CPU, RAM, SSD storage, and data transfer, and it adjusts burst prices every few minutes [5]. Amazon Web Services (AWS) [1], Azure [4], and Google Cloud Platform [6] all offer a pay-as-you-go pricing method. AWS Lambda [2] and Azure Functions [3] allow uploading code and paying for computing time only when the code is triggered to run.

RaaS systems use economic mechanisms, such as auctions, to allocate resources [14,23,45]. AWS EC2 spot instances [12], Alibaba Cloud spot instances [7], and Packet spot market [8] are examples of auctions in horizontal

© Springer Nature Switzerland AG 2019
M. Coppola et al. (Eds.): GECON 2018, LNCS 11113, pp. 24–38, 2019.
https://doi.org/10.1007/978-3-030-13342-9_3

elasticity. We predict that auctions will be deployed in vertical elasticity as they are now deployed in horizontal elasticity.

The Vickrey-Clarke-Groves (VCG) auction [21,25,44] is well-suited for this purpose. VCG auctions maximize the social welfare: the aggregate valuation of all users of the resource allocated to them. They allocate resources first to the users who value them most, and thus enable getting the most out of the machine. VCG auctions are already used by Facebook to allocate ad spaces [33] and have been used by Google for contextual ads since 2012 [43]. They have also been shown suitable for network bandwidth allocation [28].

Ginseng systems are examples of VCG auctions used for resource allocation (for RAM [14] or cache [23]). In this work we focus on Ginseng for RAM (referred to as Ginseng from this point on). Ginseng [14] consists of a market-driven RAM allocation mechanism called Memory Progressive Second Price (MPSP) that resembles a VCG auction. In Ginseng, guests run economic agents who bid for resources auctioned by the host. Thus, the mechanism incentivizes selfish, rational agents, who only care about their own profit, to bid with their true valuation of the RAM. Ginseng was shown to achieve up to 16% improvement in guest benefit from RAM allocation and up to 43% improvement from cache allocation, compared to state-of-the-art approaches. Since the number of auction participants is bound by the number of VMs on a single physical machine, the host computation time is not a bottleneck. For example, for 24 guests, the computation takes less than a second.

However, VCG auctions are not perfect. To maximize the social welfare, they incur additional costs to the users, possibly hindering profitability. VCG auctions, especially in repeated settings, are also not collusion-proof [18,19,22,24,29,31,37,39,41,42]. Colluding to increase profit may reduce social welfare [22], e.g., by bid rotation [16,17,32,36] or sub-optimal redistribution of the goods [22,37]. In a cloud environment, goods can only be transferred with the host's consent, so a collusion scheme involving resource transfer is impossible, but other forms of collusion are possible. For example, consider a cloud machine with 10 GB of RAM, and two VMs running memory-heavy applications, requiring the entire available RAM. Suppose one VM owner (Alice) values the RAM at 15¢ per hour, while the other (Bob) values it at 10¢ per hour. Alice and Bob can discover who the other VM belongs to [38], and they can agree on a collusion scheme. Since agents care only about their profits and are not exposed to each other's private information, there is no guarantee that they will agree on an efficient scheme, where Alice gets the RAM. Instead, they might agree that Bob gets the RAM and compensates Alice 7¢ per hour for not bidding. Both their profits increase, but the social welfare drops from 15¢ to 10¢ per hour.

We propose a platform for collaboration among guests that will increase their profits, thus reducing their incentive to collude, without changing the auction efficiency and the social welfare. In this model, guests can ask the host to consider them as a single guest when computing their bill. Since MPSP is based on exclusion-compensation [28], where guests pay for the damage they cause others, they would pay less if they are billed together. The host acts as a trusted third party, since guests already share their (private) valuations with the host in the auction. They tell the host how they want to share the discount (in terms of profit parts). The host calculates the reduced bills accordingly. A host might support such interactions if its main revenue is from a base payment rate, as in Ginseng, where each guest rents a base amount of the resource for a fixed price. A host also might support such interactions in private clouds, where the only goal is to maximize social welfare. This mechanism does not significantly increase the computational load on the host: computing a bill for a coalition is computationally equivalent to computing the bill for a single guest.

Our contribution is the proposal, implementation, and evaluation of a guest coalition mechanism that does not harm social welfare and which RaaS hosts have an incentive to support. It allocates resources efficiently while lowering guests' costs, thus reducing their incentive to collude in a harmful manner. The economic mechanism can thus be used for its original purpose: optimal resource allocation, which leads to optimal hardware use. The implementation was released as free software [10].

2   Simplified Memory Progressive Second Price Auction

We begin by describing a representative VCG-like resource auction, called Simplified Memory Progressive Second Price (SMPSP). It resembles bandwidth auctions [28,30], and is identical to the MPSP auction used by Ginseng [14] when guests have monotonically increasing, concave functions, consistent with diminishing returns (which are common in auction schemes [28,30]). The SMPSP auction is identical to repeating the auction proposed by Maillé and Tuffin [30] when the valuation function is approximated by using a single point. SMPSP is also used by Movsowitz *et al.* [35]. As in [30], this auction converges to approximately optimal social welfare, and therefore approximately optimal allocation.

SMPSP is a repeated auction, performed in rounds, each composed of steps. In each SMPSP round, the host first announces the free amount of resource for rent. Each guest i responds with a bid: a pair (p_i, q_i), where p_i is the unit price the guest is willing to pay to rent a resource quantity q_i (the *bid quantity*). The host collects all bids. Guests are sorted in decreasing order by their unit prices and allocated their bid amount (or the free amount left). Bills are calculated according to the exclusion compensation principle. Let q'_i denote the amount allocated to guest i, and let q''_j denote the amount guest j would have received if guest i did not exist. Then the bill guest i would pay is given by

$$B_i = \frac{1}{q'_i} \sum_{j \neq i} p_j (q''_j - q'_j). \tag{1}$$

Finally, guests are notified of the new allocation and the host redistributes the resource.

Throughout this work, we assume that allocation efficiency is more important to the provider than the guest auction payments. This assumption always holds in private clouds, and sometimes in public clouds. In public clouds, the available resources usually suffice for all the guests. When there is temporary resource pressure, the auction mechanism serves as a bridge solution until the problem is solved: either by the client purchasing reserved resources, or by the provider migrating the client to a different physical machine. Therefore, in public clouds as well, the auction payments are not the main source of revenue, but auction efficiency is crucial, as it allows the provider more time to handle the migration.

We focus on the case where the resource does not suffice for all the guests, as sometimes happens in spot instances [1,7,8]. Otherwise, SMPSP bills are 0 and the coalitions are unnecessary.

3 Related Work

In VCG and VCG-like auctions, guests may collude in various ways, most of which reduce the social welfare by preventing optimal allocation. One possible collusion scheme is a secondary resource market, conducted without the host's awareness [24,29,32]. In this scheme, a bidding ring is formed: a special trusted agent or third party acts as the ring center and holds a knockout auction among ring members to decide on the winner and on money transfers. In another work [18] the ring center is replaced by complete knowledge of the ring members' valuations, and the designated winner passes the won goods to other colluders. Neither the existence of a trusted specific agent nor the complete knowledge of valuations is practical in our case, due to the lack of trust between guests. In addition, in our model goods can only be transferred with the host's consent.

Another possible scheme is price shading, where one agent bribes another for falsely reporting its bid price or reducing it to zero [22,37]. This scheme can harm the allocation efficiency: there is no guarantee that the briber is the agent with the higher valuation, and, when bills are artificially lowered, winning guests are incentivized to increase their bid quantity beyond their original request (without collusion). Since the original allocation was approximately optimal, the change will harm its efficiency.

Bid rotation [16,17,19,32,36,41] can be seen as a private case of price shading in repeated auctions, where agents take turns falsely reducing their bid price to 0 (effectively not participating in the auction). In addition to harming the social welfare, when the auctioned resource is RAM, this scheme also harms the agents themselves: frequently passing RAM from one agent to another reduces its utility, as demonstrated by Movsowitz et al. [35].

Kraus et al. [27] present coalition formation in a task oriented domain: agents form a group to perform a task and must decide on a division of the profit. Several division strategies are proposed: equal division of profits, division proportional to contribution, or by an algorithm that finds a stable division—such that there is no group of agents with an incentive to defect. An agent can choose to forgo some profit in each of the division strategies. Simulations showed that equal division with a compromise of 20% yields the highest profits. Implementing these findings in our model is left for future work. Chatterjee et al. [20] present bidding rings as a coalition formation bargaining game, but assume bidders have complete information regarding their valuations.

The negative effects of collusion between auction participants and the recommended auctioneer responses have been widely studied. Some works study the effects of money transfers [32], cooperation enforcement types [31], and auction types [31, 42] on viable collusion schemes. Second price auctions were found to be more amenable to incentive-compatible collusion schemes than first price. Recommended auctioneer responses include reserve prices [24, 32], ceiling prices [26], and withholding information about the value of the goods [39].

Movsowitz *et al.* [34] review cloud attacks that can be applied in the RaaS model and various ways to defend against them. Movsowitz *et al.* [35] study economic attacks that are unique to the RaaS model, taking advantage of the economic mechanisms used in RaaS clouds.

Our collaboration mechanism does not involve agents changing their bids and does not assume any trust between them. If the attackers are rational, i.e., profit-driven, our mechanism makes both attacks and collusion less appealing, by reducing the profit to be gained from them.

4   Negotiation Protocol

Agents form a coalition that the host considers as a single agent when computing their bill. Agents are still charged separately, but the sum of their bills is lower. They negotiate the division of the additional profit.

The agents negotiate after each auction round's results are announced, before submitting a new bid (see Fig. 1). Each agent can negotiate with all others, but can only create a coalition with one other agent/coalition in each round. This is a design choice intended to simplify the protocol. It could easily be changed so that each auction round contains multiple negotiation rounds.

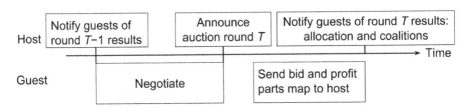

Fig. 1. Coalition protocol: Auction timeline

4.1   Negotiation Between Two Agents

In the base case of the protocol, two agents negotiate the creation of a coalition. Each protocol step consists of an offer, followed by a positive response (*accept*), a rejection or a counter-offer. An offer is a map from agents to *profit parts*—positive fractions of the profit each agent receives, whose sum is 1.

If a coalition is formed, both agents send the host this map. The host uses it to compute their bills as follows: Let T_C denote the bill for the entire coalition, calculated as if agents in the coalition were one combined agent. If agent i gets amount q_i' of the resource, let $B_i(q_i')$ denote i's undiscounted bill for it. Finally, let f_i denote the fraction of the coalition's discount that i should get. Then the new bill for coalition member i is

$$ bill_i = B_i(q_i') - f_i \left(\sum_{j \in C} B_j(q_j') - T_C \right). \tag{2} $$

An agent's bill may be negative. For example, if i had no bill before joining the coalition, i.e., $B_i = 0$, and i and j are the only agents in the system, then the total bill of their coalition is 0. But since i has a positive profit part, the host transfers money from j to i.

Upon notification of an auction round's results, the host also notifies each agent of its actual bill, the bill it would have paid outside the coalition, and the profit parts map (to confirm the coalition's validity). The agent can then compute its profit from the coalition in this round.

4.2   Negotiating a Coalition of Coalitions

In the following rounds, a two-agent coalition can grow by negotiating with another agent/coalition (Fig. 2). The leader of the larger of the two coalitions leads the joint coalition in further negotiations. If coalition sizes are equal, the agent whose proposal was accepted is the leader. If approached, a non-leader defers to the leader. Only the agents are aware of the chosen leader: this role is irrelevant for the host.

This design decision encourages coalition growth, since the leader of the larger coalition has successfully negotiated before. Moreover, if only agents that were allocated their full bid quantity are allowed to participate in coalitions (see Sect. 5.1), that leader is likely to be allowed to participate in the next auction round as well. For same-size coalitions, choosing the accepting party as the leader may mean choosing a leader whose proposals have been rejected, and thus are likely to be rejected in the future, preventing the coalition's growth.

Two leaders negotiate the merger of their respective coalitions similarly to the two-agent protocol. First they report their coalition sizes in a preliminary protocol step. Leaders normalize the sent or received offers by coalition sizes (as explained below). After negotiating the profit parts as in the two-agent case (see Sect. 5.2), each leader communicates its profit parts map to the other leader and informs its old coalition members of the negotiation results and the new leader, so that every member can compute the new map and send it to the host.

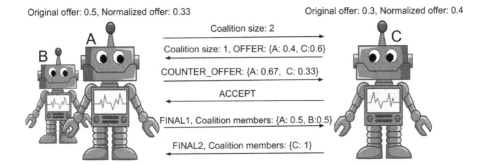

Fig. 2. Coalition protocol: Two leaders negotiating a merge.

4.3 Normalization of Offer Values

To prevent the order in which agents join a coalition from devaluing profit parts, leaders normalize the offer values. If a single agent A would have offered $0 < \alpha < 1$ to a potential single partner B, then when agents A and B lead coalitions C_A and C_B respectively, A offers B $2\alpha \frac{|C_B|}{|C_A|+|C_B|}$. As long as $0 < \alpha \leq \frac{1}{2}$, the normalized value is in $(0, 1)$. For two coalitions of size 1, the normalized value equals α.

4.4 Leaving a Coalition

An agent who wishes to leave a coalition need only refrain from sending the map to the host. The host will exclude this agent from the coalition, normalize the other members' profit parts, and notify the other members. If the leader leaves the coalition, the coalition dissolves, and members must negotiate to create new ones.

4.5 Finding the Majority Version

The host may receive different map versions for the same coalition, e.g., when an agent wants to leave the coalition or dishonestly try to divide the profit parts differently. In this case, if there is a majority version, the host will consider it the true map. The host will remove the agents who sent a different map, normalize the map, and notify the remaining members. Otherwise (there is no majority version), the coalition is rejected, and an empty map is returned.

5 Host Policy and Agent Strategy

In view of the suggested protocol, the host and agents may use various strategies. In this paper we are interested in identifying plausible host policies which lead to our goal of stable and social-welfare-safe collaborations. To evaluate how these policies actually lead to our goals, we analyze how guests should behave under these policies to optimize their profit. This allows us to accurately model the behavior of plausible, rational guests and justify our conclusions regarding the proposed protocol.

5.1 Host Policy

In addition to collecting the agents' coalition requests and returning the resulting map, the host may enforce rules on agents' participation in coalitions.

To preserve social welfare, the host should enforce a **no-increase policy**: an agent who attempts to increase a bid quantity will be removed from the coalition. We will demonstrate how without this policy, allocation efficiency is compromised. Agents evaluate their *utility* from a possible bid quantity q as $U(q) = V(q) - bill(q)$, where $V(q)$ is their valuation and $bill(q)$ is the estimated bill they will pay. This estimation is based on previous bills. Let i denote such an agent. Since i's valuation function V_i is monotonically increasing and concave, i's participation in a coalition increases $argmax_q(U)$ and as a result—i's bid quantity. Unless i is the last guest to be allocated a resource, this will cause at least one agent to get less of the resource. The original allocation is based on truthful bids and therefore is approximately optimal (as in [28]), so any deviation from it reduces the social welfare.

The no-increase policy does not impede the system's ability to accommodate changes in the guest valuation function over time. If the load on the guest increases, it will be worthwhile to the agent to increase its bid quantity and leave the coalition.

A host may decide that only agents allocated their full bid quantity may be admitted to a coalition. We define such agents as **satisfied**. Allowing only satisfied agents in a coalition protects the host from a possible strategy of unsatisfied agents: if they can join a coalition, they can attempt to increase their compensation by pretending that an allocation causes them more harm than it actually does. They do so by bidding for a large quantity in advance, knowing they will neither win it nor pay for it but will still get compensated by other coalition members.

With these precautions, the social welfare is likely to be preserved. However, the mechanism is still vulnerable to agents leaving the coalition, increasing their resource demand, and then rejoining a coalition. Allowing only satisfied guests in coalitions reduces the potential profit of this scheme, but does not eliminate it completely.

5.2 Agent Strategy

Given the protocol we defined and the host's policy, what should the agent's strategy be?

Bill Estimation. SMPSP agents are *p-truthful*: for a bid quantity q_i, and valuation function V, the best strategy for choosing a bid price is $p_i(q_i) = \frac{V(q_i+base)-V(base)}{q_i}$ [14]. So, agents need only decide on the bid quantity.

A rational agent in a coalition should weigh the consequences of increasing q_i, assuming its bill is undiscounted for the new amount, at least until rejoining a coalition. Let $D_i(q_i)$ denote agent i's estimated discount from the coalition. i estimates it will take N_1 rounds to rejoin a similar coalition, and N_2 rounds for the game to end or for i's valuation to change. Finally, let q_i^{prev} denote i's previous bid quantity. So, when considering a bid quantity q_i, i will estimate the expected value of its future bill as:

$$bill_{est}(q_i) = \begin{cases} B_i(q_i) - D_i(q_i), & \text{if } q_i \leq q_i^{prev} \\ B_i(q_i) - D_i(q_i)\frac{N_2-N_1}{N_2}, & \text{otherwise} \end{cases} \tag{3}$$

In other words, the agent expects to lose the discount (due to the host's no-increase policy) until rejoining the coalition after N_1 rounds. However, if the agent does not increase q, then it expects to stay in the coalition and continue paying the discounted price. This means that agents will avoid changing their bid quantity unless necessary, for example when their valuation changes.

Agent i can estimate $B_i(q_i)$ and $D_i(q_i)$ on the basis of i's previous bills and discounts. Agent i can estimate the time during which its own load and valuation are expected to remain the same. When it expects its resource requirements to change, it also expects the coalition to break. In addition, i may learn the typical duration of a coalition, according to its experience with specific partners. N_1 can be estimated optimistically as $log_2(|coalition|)$, as would happen if all possible coalitions are formed in each round, or pessimistically as $|coalition|-1$, as would happen if a single guest joins the coalition in each round. N_1 and $D_i(q_i)$ also depend on the agent's negotiation strategy.

Negotiation Strategy. The negotiation strategy of an agent is a pair (OV, AT), where Offer Value $OV \in (0,1)$ is the initial profit part offer the agent proposes, and Accept Threshold $AT \in (0,1)$ is the minimal profit part the agent is willing to accept. A high AT and low OV will increase the agent profit $(D_i(q_i))$ in a given coalition. However, such values will discourage potential partners from forming a coalition with the agent, i.e., increase N_1. Suppose the agent has learned the others' values, e.g., from previous negotiations. Then to increase its chances of joining a coalition sooner rather than later, the agent should choose a high OV, just above the others' values, and a suitably low AT. This will maximize the number of rounds the agent spends in coalitions and the total profit over time (a formal analysis is available in the technical report [9]). Hence, a uniform environment, where all agents have the same (OV, AT), is unstable. We focus instead on a mixed environment, where each agent has its own negotiation profile, and we conduct experiments to find the optimal values.

6   Experimental Evaluation

We experimented to find good strategies in a mixed environment, to determine the effect of coalitions on agent profits and social welfare, and to characterize the connection between coalition size and agent profits.

6.1   Experimental Setup

Experiments were run on a server with 24 cores and 16 GB of RAM, running Ubuntu Linux 4.4.0-53-generic. Agents were run on virtual machines in an implementation of Ginseng [14]. The code was released as free software and is available from [10].

Our experimental environment represents a cloud machine, with 10 VMs running servers. Each server runs elastic memcached [14]—an elastic memory version of a widely used key-value cloud application. Memcached has a concave performance function. For the valuation function we used $V(performance) = c \cdot performance$. c is a constant, chosen for each guest i.i.d. from a Pareto distribution with $\alpha = 1.36$, as in [14]. The available RAM for auction was determined as one-third of the aggregate demand, to create resource pressure, under which there are payments in the SMPSP auction.

6.2   Optimal Negotiation Profile

We restricted the agent strategy to choosing a constant negotiation profile and analyzed which profiles yield the most profit. Analyzing strategies which evolve over time and depend on partner identity is left for future work.

To create a mixed environment, we sampled 20 valuation sets and ran 5 experiments of 40 rounds each for each set. In each experiment OV and AT were drawn i.i.d. from a uniform distribution. $OV \sim U(0,1)$, and $AT \sim U(0, 1 - OV)$, since an agent should not decline a value above what would be obtained from offering OV. If after normalization (Sect. 4.3) $OV > 1$, then the agent would refrain from offering it. All agents were allowed to participate in coalitions. Agents' valuations were shuffled every 10 rounds so that coalitions would break (due to the no-increase policy), allowing for more negotiation opportunities.

In a mixed environment, accept thresholds matter. Too high a threshold means being too "picky". Too low a threshold means settling for deals with lower profits. Offer values matter as well. When the offer value is low, the agent keeps more of the profit, and earns more from each coalition. When it is high, so is the agent's chance of joining a coalition. In our settings, the optimal AT is around 0.8 (Fig. 3a) and the optimal OV is around 0.2 (Fig. 3b). The profit peak at $OV = 0.2$ is caused by the way ATs are drawn: over half of the ATs drawn this way are below 0.2. Therefore, if an agent's OV is higher than 0.2, most partners are likely to accept it. As OV grows, the agent relinquishes a larger part of the profit to its partner, without notably increasing the chances of acceptance while decreasing profits. Exact values depend on the profile distribution.

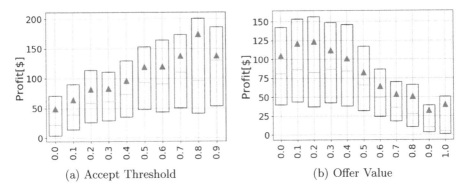

Fig. 3. Profit from coalitions as a function of negotiation parameters in a mixed environment. Mean values are shown as triangles. The boxes represent the first, second and third quartiles.

6.3   Social Welfare and Guest Profit

How are the social welfare and profit affected by negotiations? Given our results in Sect. 6.2, we chose a symmetric profile that is likely for rational agents: $OV = 0.2$ and AT just below it at 0.15, so that coalitions can form. With this profile we ran three experiment scenarios: coalitions are allowed for satisfied agents, coalitions are allowed for all agents, or coalitions are prohibited. Each experiment ran for 40 rounds. Valuations for the guests are as in Sect. 6.2, but they were not shuffled.

In each experiment, we calculated the social welfare and the mean profit for guests in each round. The profit is the difference between the valuation of the resource and the actual (occasionally discounted) amount paid. While the social welfare remains the same (i.e., the allocation of goods remains efficient), the mean profit increases by up to 67% (60% on average), as shown in Fig. 4a. When coalitions are allowed for all agents, the mean profit reaches the profit from the optimal collusion scheme, where guests share their entire private valuations and bid for the same quantities they would get from the host, so their bill is zero.

6.4   Optimal Coalition Size

What is the optimal coalition size for agents? We ran two experiment batches, one with only satisfied guests and one with all guests. In each batch, 5 different valuation sets were drawn. Each set was run 10 times.

Although a single coalition of all agents has the largest aggregate profit, it does not necessarily optimize the mean profit of a single member. In fact, the opposite is true: some agents, when joining the coalition, do not increase the total coalition profit but share it: their mean profit does not monotonically increase with the coalition size, (see Fig. 4b). This result holds whether or not unsatisfied agents are allowed. It means that a single coalition of all agents is not

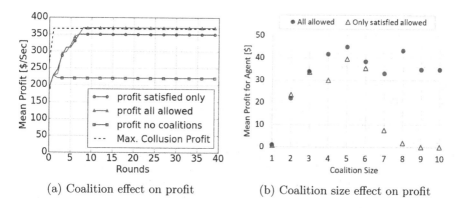

(a) Coalition effect on profit   (b) Coalition size effect on profit

Fig. 4. (a) shows the mean profit for guests in each round, for different host coalition support policies. (b) shows the mean profit from coalitions per agent in a single round as a function of the coalition size.

in the agents' best interest, and so it is unstable. The instability can be resolved using a Shapley value based division, where each agent's profit depends on its contribution.

7   Conclusion and Future Work

As cloud computing advances, we expect more clouds to shift to economic mechanisms such as auctions for selling cloud resources. As they do so, the importance of understanding and handling collusion between auction participants will grow. We presented and implemented a negotiation mechanism for coalition formation in a VCG-like auction, which is suitable for auctioning computing resources. It maintains the auction's social welfare while reducing the participants' costs. Although coalition participants pay lower bills, they nonetheless refrain from unnecessarily increasing their resource demands. In a mixed environment, profit will be maximized for an agent willing to accept a profit share of at least 45% and who offers partners 40%. Nor do large coalitions necessarily benefit their members. We conjecture that negotiation strategies that change over time and with different partners may increase the agents' profit. Exploring different division schemes such as Shapley [40]-based division is left for future work.

VCG auctions can also be used for multi-resource allocation using the same exclusion-compensation principle as for a single resource. Therefore, given a VCG multi-resource auction implementation, the anti-collusion coalition mechanism can be applied to it as well.

Acknowledgments. This work was partially funded by the Amnon Pazi memorial research foundation. We thank D. Parkes for fruitful discussions. We thank A. Bousso, Y. Lev, A. Ohayon, and S. Levenzon Kuninin for their contributions to the software implementation.

References

1. Amazon EC2 pricing. https://aws.amazon.com/ec2/pricing/
2. AWS Lambda. https://aws.amazon.com/lambda/
3. Azure functions pricing. https://azure.microsoft.com/en-us/pricing/details/functions/
4. Azure pricing. https://azure.microsoft.com/en-us/pricing/
5. CloudSigma price schedules: Burst pricing. http://cloudsigma-docs.readthedocs.io/en/2.14/billing.html#burst-levels
6. GCP pricing—google cloud platform. https://cloud.google.com/pricing/
7. Spot instances - product introduction—alibaba cloud documentation center. https://www.alibabacloud.com/help/doc-detail/52088.htm
8. Spot marketing pricing - discount packet bare metal servers. https://www.packet.net/bare-metal/deploy/spot/
9. Agmon, S., Agmon Ben-Yehuda, O., Schuster, A.: Preventing collusion in cloud computing auctions. Technical report CS-2018-01, Technion (2018). http://www.cs.technion.ac.il/users/wwwb/cgi-bin/tr-info.cgi/2018/CS/CS-2018-01
10. Agmon, S., Kuninin, S., Bousso, A., Lavi, N.: RaaS negotiations (2018). https://bitbucket.org/shunit/negotiations
11. Agmon Ben-Yehuda, O., Ben-Yehuda, M., Schuster, A., Tsafrir, D.: The Resource-as-a-Service (RaaS) cloud. In: Proceedings of the 4th USENIX Conference on Hot Topics in Cloud Computing, p. 12. USENIX Association (2012)
12. Agmon Ben-Yehuda, O., Ben-Yehuda, M., Schuster, A., Tsafrir, D.: Deconstructing Amazon EC2 spot instance pricing. ACM Trans. Econ. Comput. **1**(3), 16 (2013)
13. Agmon Ben-Yehuda, O., Ben-Yehuda, M., Schuster, A., Tsafrir, D.: The rise of RaaS: the Resource-as-a-Service cloud. Commun. ACM **57**(7), 76–84 (2014)
14. Agmon Ben-Yehuda, O., Posener, E., Ben-Yehuda, M., Schuster, A., Mu'alem, A.: Ginseng: market-driven memory allocation. SIGPLAN Not. **49**(7), 41–52 (2014)
15. Agmon Ben-Yehuda, O., Schuster, A., Sharov, A., Silberstein, M., Iosup, A.: Expert: Pareto-efficient task replication on grids and a cloud. In: 26th IEEE International Parallel and Distributed Processing Symposium, IPDPS, pp. 167–178 (2012)
16. Aoyagi, M.: Bid rotation and collusion in repeated auctions. J. Econ. Theory **112**(1), 79–105 (2003)
17. Aoyagi, M.: Efficient collusion in repeated auctions with communication. J. Econ. Theory **134**(1), 61–92 (2007)
18. Bachrach, Y.: Honor among thieves: collusion in multi-unit auctions. In: Proceedings of the 9th International Conference on Autonomous Agents and Multiagent Systems: Volume 1, pp. 617–624. International Foundation for Autonomous Agents and Multiagent Systems (2010)
19. Blume, A., Heidhues, P.: Modeling tacit collusion in auctions. J. Inst. Theor. Econ. JITE **164**(1), 163–184 (2008)
20. Chatterjee, K., Mitra, M., Mukherjee, C.: Bidding rings: a bargaining approach. Games Econ. Behav. **103**, 67–82 (2017)
21. Clarke, E.H.: Multipart pricing of public goods. Public Choice **11**(1), 17–33 (1971)
22. Eső, P., Schummer, J.: Bribing and signaling in second price auctions. Games Econ. Behav. **47**(2), 299–324 (2004)
23. Funaro, L., Agmon Ben-Yehuda, O., Schuster, A.: Ginseng: market-driven LLC allocation. In: 2016 USENIX Annual Technical Conference, pp. 295–308. USENIX Association, Berkeley (2016)

24. Graham, D.A., Marshall, R.C.: Collusive bidder behavior at single-object second-price and English auctions. J. Polit. Econ. **95**(6), 1217–1239 (1987)
25. Groves, T.: Incentives in teams. Econometrica **41**(4), 617–631 (1973)
26. Johnson, P., Robert, J., et al.: Collusion in a model of repeated auctions. Université de Montréal, Centre de recherche et développement en économique (1999)
27. Kraus, S., Shehory, O., Taase, G.: The advantages of compromising in coalition formation with incomplete information. In: Proceedings of the Third International Joint Conference on Autonomous Agents and Multiagent Systems-Volume 2, pp. 588–595. IEEE Computer Society (2004)
28. Lazar, A., Semret, N.: Design and analysis of the progressive second price auction for network bandwidth sharing. Telecommun. Syst. Spec. Issue Netw. Econ. **20**, 255–263 (1999)
29. Mailath, G.J., Zemsky, P.: Collusion in second price auctions with heterogeneous bidders. Games Econ. Behav. **3**(4), 467–486 (1991)
30. Maillé, P., Tuffin, B.: Multi-bid auctions for bandwidth allocation in communication networks. In: IEEE INFOCOM (2004)
31. Marshall, R.C., Marx, L.M.: Bidder collusion. J. Econ. Theory **133**(1), 374–402 (2007)
32. McAfee, R.P., McMillan, J.: Bidding rings. Am. Econ. Rev. **82**(3), 579–599 (1992)
33. Metz, C.: Facebook doesn't make as much money as it could - on purpose (2015). https://www.wired.com/2015/09/facebook-doesnt-make-much-money-couldon-purpose
34. Movsowitz, D., Agmon Ben-Yehuda, O., Schuster, A.: Attacks in the Resource-as-a-Service (RaaS) cloud context. In: Bjørner, N., Prasad, S., Parida, L. (eds.) ICDCIT 2016. LNCS, vol. 9581, pp. 10–18. Springer, Cham (2016). https://doi.org/10.1007/978-3-319-28034-9_2
35. Movsowitz, D., Funaro, L., Agmon, S., Agmon Ben-Yehuda, O., Dunkelman, O.: Why are repeated auctions in RaaS clouds risky? In: Coppola, M., Carlini, E., D'Agostino, D., Altmann, J., Bañares, J.Á. (eds.) GECON 2018. LNCS, vol. 11113, pp. 39–51. Springer, Cham (2019)
36. Rachmilevitch, S.: Endogenous bid rotation in repeated auctions. J. Econ. Theory **148**(4), 1714–1725 (2013)
37. Rachmilevitch, S.: Bribing in second-price auctions. Games Econ. Behav. **92**, 191–205 (2015)
38. Ristenpart, T., Tromer, E., Shacham, H., Savage, S.: Hey, you, get off of my cloud: exploring information leakage in third-party compute clouds. In: Proceedings of the 16th ACM Conference on Computer and Communications Security, pp. 199–212. ACM (2009)
39. Seres, G.: Auction cartels and the absence of efficient communication. Int. J. Ind. Organ. **52**, 282–306 (2017)
40. Shapley, L.: Stochastic games. Proc. Natl. Acad. Sci. USA **39**(10), 1095–1100 (1953)
41. Skrzypacz, A., Hopenhayn, H.: Tacit collusion in repeated auctions. J. Econ. Theory **114**(1), 153–169 (2004)
42. von Ungern-Sternberg, T.: Cartel stability in sealed bid second price auctions. J. Ind. Econ. **36**(3), 351–358 (1988)

43. Varian, H.R., Harris, C.: The VCG auction in theory and practice. Am. Econ. Rev. **104**(5), 442–45 (2014)
44. Vickrey, W.: Counterspeculation, auctions, and competitive sealed tenders. J. Finance **16**(1), 8–37 (1961)
45. Yu, D., Mai, L., Arianfar, S., Fonseca, R., Krieger, O., Oran, D.: Towards a network marketplace in a cloud. In: HotCloud (2016)

Why Are Repeated Auctions in RaaS Clouds Risky?

Danielle Movsowitz[1][(✉)], Liran Funaro[2], Shunit Agmon[2],
Orna Agmon Ben-Yehuda[2,3], and Orr Dunkelman[1]

[1] Computer Science Department, University of Haifa, Haifa, Israel
`dmovsowi@campus.haifa.ac.il,orrd@cs.haifa.ac.il`
[2] Computer Science Department, Technion—Israel Institute of Technology,
Haifa, Israel
`{funaro,shunita,ladypine}@cs.technion.ac.il`
[3] Caesarea Rothschild Institute for Interdisciplinary Applications of Computer
Science, University of Haifa, Haifa, Israel

Abstract. The world of cloud computing is progressing from the concept of securing resources by predefined units to dynamically allocating resources using economic mechanisms. New mechanisms offer better utilization of the hardware by sharing it among multiple users. However, they allow new types of economic attacks. We introduce two new economic attacks performed by malicious users. These attacks harm the aggregate utility of Resource-as-a-Service (RaaS) clouds. Our first attack aims at raising bills in the system, and causing victims to pay more for the same amount of resources. Over time the attack may cause victims to exhaust their budget, thus lowering their demand for resource allocation, and allowing the attacker to acquire the freed resources at a negligible cost. Our second attack is designed to hinder the victim's performance at specific points in time by outbidding them for a single round. For resources of high regaining costs or that their full utilization takes time (e.g., RAM), even a single round without the resource may significantly hinder the performance. In this work we demonstrate on a simple representative example how the first attack reduces the victim's profit sevenfold and the second attack causes damage of \$290–\$630 for every dollar spent on the attack.

Keywords: VCG · Resource allocation · RaaS · Economic attacks

1 Introduction

The Resource-as-a-Service (RaaS) cloud [3] is an economic model of cloud computing that allows providers to sell adjustable quantities of individual resources (such as CPU, RAM, and I/O resources) for short intervals—even at a sub-second granularity. In the RaaS cloud, clients can purchase exactly the resources they need when they need them. As price wars drive cloud providers towards this model [5], they start offering plans for dealing with resource requirement bursts:

© Springer Nature Switzerland AG 2019
M. Coppola et al. (Eds.): GECON 2018, LNCS 11113, pp. 39–51, 2019.
https://doi.org/10.1007/978-3-030-13342-9_4

CloudSigma offered time-varying burst prices in 2010 [11], Amazon EC2 offered burstable performance instances in 2014 [14], Google Cloud offered Pay-as-you-go in 2016 [17], and Microsoft introduced the burstable Azure cloud Instance in 2017 [8]. When resources are dynamically rented, e-commerce requires calculating online economic decisions. Such decisions can only be made in real time by automated agents. E-commerce also requires efficient and computationally simple allocation mechanisms. These mechanisms may be centralized (as in an auction) or decentralized (as in a marketplace [37] or by negotiations [2]).

We see that horizontal scaling (adding more machines) has already matured to the point of incorporating advanced economic mechanisms such as auctions (e.g., AWS Spot Instances [4], Packet [31], and Alibaba Cloud Spot Instances [7]). Nevertheless, in the case of vertical scaling (increasing an existing machine's resources) we are only now seeing signs of early adoption of such mechanisms (e.g., Amazon EC2 T2 Instances, Google Cloud Platform, CloudSigma).

In the past few years, numerous studies have been published regarding different attack methods relevant to clouds, e.g., side channel [30], Resource Freeing Attack (RFA) [33], co-location attacks [34], and Economic Denial of Sustainability (EDoS) [20]. Most of the studied attacks are aimed at penetrating the security of the system and not at the economic mechanism that drives the resource allocation in the system. EDoS attacks are an exception: they cause victims to scale their resources beyond their economic means. In this work we take this line of vulnerabilities further, presenting combined economic-computer-science attacks.

Our contribution is the design of two low-cost economic attacks aimed at auction based clouds. The implementation and evaluation of the attacks were done on a simple representative example using Ginseng [6], a market driven cloud system for efficient RAM allocation. The first attack is the Price Raising attack. This attack raises prices in the system, thus reducing the victim's profit and forcing it to free resources. This enables the attacker to rent the freed resources at a negligible cost. The second attack is the Elbowing attack. This attack hinders the victim's performance by outbidding it for a single round at specific points in time. Due to the nature of RAM usage, the victim suffers from reduced performance even after the attack round ends, and it re-acquires the RAM. We demonstrate how the *Price Raising attack* reduces the victim's profit sevenfold and the *Elbowing attack* causes damage of $290–$630 for every dollar spent on the attack.

In Sect. 2 we describe the auction protocol we attack. In Sect. 3 we discuss the vulnerabilities in repeated auctions, and the motivations behind attacking such auctions. In Sect. 4 we describe the experimental setup. In Sect. 5 we present and analyze our first attack—the Price Raising attack, and in Sect. 6 we present and analyze our second attack—the Elbowing attack. We review related work in Sect. 7, and conclude in Sect. 8.

2   Background: An Auction Mechanism for Vertical Scaling in the Cloud

In resource auctions, guests have private valuations for each resource (e.g., RAM or bandwidth) that reflect how much additional resources are worth to them. In a full Vickrey-Clarke-Groves (VCG) auction [12,18,35], guests bid with a full valuation function. The auctioneer chooses the allocation which maximizes the aggregate valuation of the allocated resources (the social welfare). The social welfare is defined as the sum of all of the guests' valuation for resources in a specific allocation. The guests pay according to the exclusion compensation principle: they pay the difference to the other guests' social welfare, incurred by their participation in the auction. We note that in the full VCG auction, guests are truthful—it is rational for them to bid with their true valuations. The Facebook's ad auction [28] and the Ginseng cache auction [16] are implementations of a full VCG resource auction.

VCG is computationally intensive, limiting its use in repeated auctions. Therefore, companies offering Spot Instances (e.g., Amazon EC2, Alibaba Cloud, and Packet), which require a repeated auction, approximate VCG using a uniform price auction. Likewise, auctioning resources in a fine granularity requires VCG approximations, which are only approximately truthful. Lazar and Semret allocate bandwidth using the Progressive-Second-Price (PSP) auction [25]. Ginseng RAM auctions RAM to guests repeatedly and frequently using the Memory-Progressive-Second-Price (MPSP) auction [6]. In this work we analyze the Simplified Memory-Progressive-Second-Price (SMPSP) auction, which resembles both these auctions. The SMPSP auction is identical to repeating the auction proposed by Maillé and Tuffin [26] when the valuation function is approximated using a single point. The SMPSP auction is also used by Agmon et al. [2]. We present the auction in detail in Sect. 2.1.

2.1   The Simplified Memory-Progressive-Second-Price (SMPSP) Auction

In a RAM auctioning system the host auctions RAM to guests. Each guest rents a permanent amount of base RAM on a constant hourly fee. In addition, the guest may participate in the re-occurring auctions to rent additional RAM.[1] The price paid in the auction for additional RAM does not affect the cost of the base RAM. The host represents the provider, and is in charge of running the auction. In order to attract more guests, the host allocates the additional RAM between the guests, using a repeated auction mechanisms, in a manner that optimizes their social welfare. Each auction round is composed of several stages:

1. **Auction Announcement**—The host announces Q, the amount of spare RAM that is auctioned in that round.

---

[1] Most of the provider's revenue comes from the constant hourly fee the guests pay for their base RAM. This allows the provider to use an auction to optimize the social welfare of the guests without worrying about its own revenue.

2. **Bidding**—Interested guests bid for a desired amount of RAM. Guest i's bid is a tuple (p_i, q_i). p_i is the maximal unit price that guest i is willing to pay for RAM (in terms of cents per Mb per hour), and q_i is the maximal quantity it is willing to accept.
3. **Allocation and Payments**—The host computes the allocation and payment according to VCG's exclusion compensation principle.
4. **Informing Guests**—The host informs each guest i of its personal results (p'_i, q'_i). The host also announces the unit price of borderline bids: the lowest accepted unit price (denoted P_{min_in}) and the highest rejected unit-price (denoted P_{max_out}). If any guest received a partial allocation, P_{min_in} is equal to P_{max_out}.

The host informs the guests about the borderline bids for three reasons. *First*, benign guests use this information to plan their next bids. They can use it to learn the minimal price they can bid with and still win some RAM. *Second*, guests can trivially learn this information over time through the rejection or acceptance of their bids, so it is futile to try and hide it. *Third*, without this information guests may bid iteratively for a few rounds to find the lowest price they can offer, thus disturbing the system's stabilization.

The results of an auction can be visually represented by a plot, as shown in Fig. 1. Each guest, in the order they were allocated RAM and sorted by the allocation algorithm, is represented by a rectangle: the width is the amount of RAM won by the guest, the height is the unit price of the guest's bid, and the rectangle's area reflects the guest's valuation for the given allocation. The guest's bill is reflected by the sum of the rectangle areas calculated from Q to q'_i. Therefore, the social welfare is given by the sum of areas that belong to guests who were allocated RAM (those in the interval $[0, Q]$).

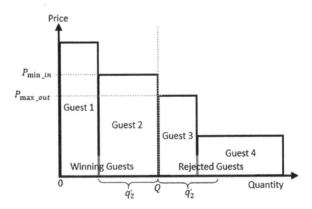

Fig. 1. Allocation plot

3 Vulnerabilities in Repeated Auctions

VCG is truthful as a single round game. The best strategy for a participant is to declare its true type (e.g., its true valuation of the resources). Even a VCG-like auction (e.g., PSP or SMPSP) is usually truthful when participants consider only a single round [6]. In the case of repeated auctions, a guest can choose a more beneficial strategy on the basis of information learned from previous bid results [9].

In the SMPSP auction protocol presented in Sect. 2.1, if a guest wants a more accurate estimation of its next bill, it can try to create a model containing the information regarding the bids of other guests in the system. This model can be created by keeping track of the data released by the host—P_{min_in} and P_{max_out}, and fusing it with the results from previous rounds—the bill paid by the guest, and the allocation it received. Given P_{min_in}, the guest knows the minimal bid that was allocated a positive quantity of RAM. At the same time, given P_{max_out} and its bill, the guest knows the maximum rejected bid price, and the average unit price in the interval $[Q, Q + q']$. The guest does not know how many bids were rejected nor their values. By changing its bid price, and following the results of each round, the guest can learn information about the rejected bids.

In Sects. 5 and 6 we will show how a guest can use the data collected in order to attack the system and gain resources for a negligible price.

3.1 Attacking a Repeated Auction Mechanism

A system based on a repeated auction is vulnerable to attacks by both adversarial guests and selfish guests wishing to improve their resource allocation. In a repeated auction, data about the system can constantly be collected. Hence, the attacker can afford sporadic non-beneficial attack rounds, as the following rounds balance out its utility.

When attacking the system, the attacker can have different direct goals driving the attack:

1. Hindering performance—The attacker prevents the victim from utilizing physical resources.
   (a) Resource deprivation—The attacker causes the victim to rent less resources.
   (b) Inefficient resource rental—The attacker harms the victim by making it suffer the overhead of re-acquiring the resource. This attack is specific for resources that have a high acquisition overhead (e.g., RAM, disk space). Even if the victim obtains access to the resource again, it would need to recover or reproduce the data. Until then, its performance is likely to be hindered.
2. Reducing profits—The attacker raises the price of resources, thus causing the victim to spend more money than planned. The immediate result is increasing the victim's expense rate, thus lowering its profit rate.

3. Reducing resource pressure (freeing resources)—If the victim's budget is drained (e.g., due to a long term cost increase), it is forced to free resources. The freed resources can then be obtained by the attacker at a lower price because the demand for the resources is lower. Even if the victim's budget is not completely depleted, it may request less resources in an effort to avoid a complete budget depletion. This reaction immediately frees resources.

An important aspect of an attack is the monetary resources required for its deployment. A high budget attack is less likely to be deployed than an attack that can be performed using a strict budget. Our proposed attacks, the *Price Raising attack* and the *Elbowing attack*, are indeed low-cost attacks. The Price Raising attack costs nothing, and the Elbowing attack causes damage of \$290–\$630 for every dollar spent by the attacker.

4   Experimental Setup

Our experiments, which verified the attacks we performed, were run on a 24-core server with 16 GB of RAM. The RAM was auctioned and allocated using Ginseng [6] over Ubunto Linux 4.4.0-72-generic. Ginseng auctions RAM to guests and then changes their physical RAM allocation accordingly. It is implemented on the KVM hypervisor [23] with Litke's memory overcommit manager MOM [1]. It controls the exact amount of allocated RAM to each guest via libvirt using ballon drivers [36]. Ginseng has a host component and a guest component. The host component includes the Auctioneer that runs the MPSP protocol that is an extension of the SMPSP protocol described in Sect. 2.1. The guest component is suitable for virtual machines (VM's).

We note that although this Ginseng implementation is based on VMs, the principals behind the attacks are also relevant to other implementation (e.g., containers).

In our experiments the victim guest machine ran an elastic version of memcached [27], a key value storage application commonly used in clouds. Elastic memcached can release the least recently used RAM, thus allowing memory usage elasticity. The performance of elastic memcached is a concave monotonically rising function of its RAM [6]. In this case, MPSP guests bid using the reduced bidding language supported by the SMPSP. Under these circumstances, the protocols are equivalent. The attacker guest machine ran MemoryConsumer [6], a synthetic application which performs linearly with its RAM consumption. The Ginseng code used to produce these results is open source, and is available from https://bitbucket.org/danimovso/ginseng-open.

5   Price Raising Attack

In the **Price Raising** attack, the attacker uses the first rounds of the auction to collect data about the borderline bids. In analogy to Dolgikh et al. [13], it learns information for free by bidding a price that is likely to be rejected. After analyzing the collected data, the attacker uses it to situate itself as the highest rejected bidder, thus setting the unit price P_{max_out}. In every auction round, the attacker requests Q—the full amount of RAM. By bidding P_{max_out} for Q, the attacker directly affects the bill paid by the winning guests (see Fig. 2). Each winner i pays $P_{max\_out} \times q'_i$. In every attack round the attacker raises P_{max_out}. This causes the winning guests to pay more for the RAM resources they receive, thus reducing their profits. The profit reduction may cause the victim to reduce its requested quantity and raise its bid price, in an effort to distance its bid from the borderline unit price, or to remain within its budget limits. Either way, the victim may free resources.

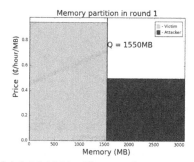
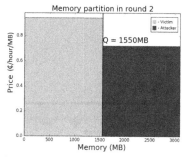

(a) Initial bids. The attacker situates itself as the guest setting P_{max_out}. Bill rate of $\$7.7/hour$ for $1.5GB$ that the victim rents.

(b) The attacker raises P_{max_out}, increasing the victim's bill rate to $\$11.16/hour$ for the same $1.5GB$.

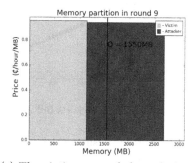

(c) The victim responds by reducing its bid quantity, and raising its bid price. The attacker gains free access to resources. The victim's bill rate is now $\$10.73/hour$ for only $1.14GB$ it rents.

Fig. 2. Price raising attack. RAM allocation plot

To keep the attack at a low budget, the attacker compares rejected guests' bids with its own valuation. If the attacker's true valuation for RAM is lower than that of any other rejected guest, then the attacker should decrease its bid price to avoid winning resources that will create additional costs that it cannot afford. Otherwise, the attacker can rent the freed RAM.

This attack achieved three goals: First, by slowly raising P_{max_out} and causing the victim to pay more for the same quantity of RAM, the attacker reduced the victim's profit sevenfold (see Fig. 3a). Second, by reducing the victim's profits and draining its budget, the attacker forced the victim to free RAM (see Fig. 3b), thus enabling the attacker to rent it at a lower cost (see Fig. 3c). Finally, by forcing the victim to rent less RAM than it ideally wanted, the attack can hinder the victim's performance.

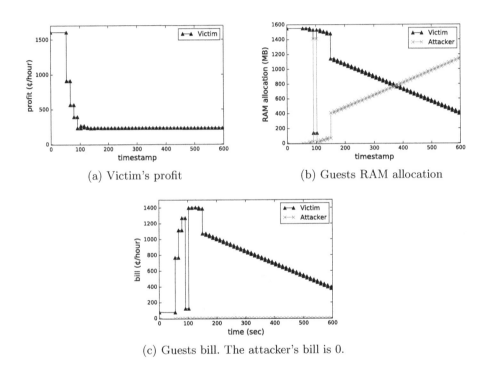

(a) Victim's profit   (b) Guests RAM allocation

(c) Guests bill. The attacker's bill is 0.

Fig. 3. Price raising attack.

6   Elbowing Attack

The **Elbowing** attack causes the degradation of the victim's performance and profit over time. This attack is designed to harm victims that fill their RAM slowly. For such victims, losing RAM even for a very short period of time can be badly damaging: while the victim re-acquires and re-fills the RAM, it loses the benefits it gained from previously owning the RAM. The attacker benefits from

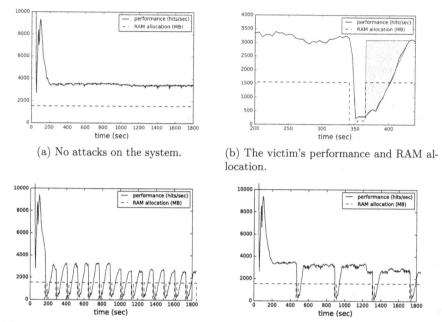

(a) No attacks on the system.

(b) The victim's performance and RAM allocation.

(c) Attack every 10 auction rounds. The victim's damage is $48/hour$ and the attack cost is $0.11/hour$.

(d) Attack every 35 auction rounds. The victim's damage is $20/hour$ and the attack cost is $0.03/hour$.

Fig. 4. Elbowing attack. The performance and RAM allocation of the victim. The victim's damage is computed using its valuation function, which attaches a value of 3 cents to every hit.

this attack since outbidding the victim every now and then does not cost nearly as much as it would cost to constantly win the allocated RAM. This means that at a relevantly low cost, the attacker can inflict severe damage to the victim.

To test this attack type we performed a parametric sweep of attacks in a system that contained two guests—the attacker and the victim. In each attack, the following process was repeated every N rounds. First, the attacker waited for the system to stabilize. During the stabilization period the victim filled its allocated RAM. Then the attacker attacked by outbidding the victim: it bid for a single round for Q with a unit price slightly higher than P_{min_in}. This forced the elastic memcached victim to lose its data (a standard, non-elastic memcached would have swapped its data to a slower storage, suffering a higher penalty). After the attack round, the attacker went back to bidding with a negligible bid price, and the victim had to re-acquire the RAM, suffering a high overhead while doing so.

We present the results of such an attack in Fig. 4. The baseline performance of the victim (without any attack) is presented in Fig. 4a. The damage from a single attack round is shown in Fig. 4b. In this case, it would be wasteful to attack again before 440 s, because although the attack only lasted between 340 s and 352 s, and the RAM was re-rented 12 s later, it was not fully filled for another 70 s after that. The shaded area in Fig. 4b represents the lingering damage of the Elbowing attack. The lingering damage is larger as the recuperation rate of the victim is slower. The longer it takes the victim to recuperate, the more cost-effective the Elbowing attack is, since the attacker does not pay extra for those rounds.

The results of examples of repeating attacks are presented in Fig. 4c and d. The attack in Fig. 4c is performed frequently (every 10 rounds), and the victim does not fully recuperate. The attack in Fig. 4d is less frequent (every 35 rounds), and thus causes more damage per single attack round. However, since the attack rate is slower, it causes smaller overall damage to the victim's profit.

The results of a parametric sweep on the attack frequency are shown in Fig. 5. Each point in the graph represents one attack, and shows the average damage caused to the victim's profit as a function of the attack average cost. The cost of the attack is determined by the frequency of the attack rounds. As shown Fig. 5 the Elbowing attack causes a damage of $290–$630 for every dollar spent on the attack. The Elbowing attacks are not necessarily low budget attacks, but they can be performed on a strict budget. To do this, the attacker needs to decide in advance how much money it is willing to spend on each attack round.

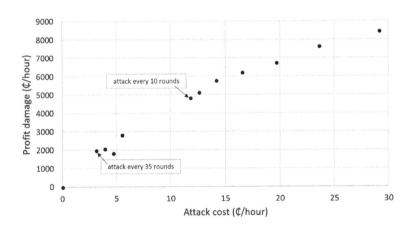

Fig. 5. Profit damage as a function of the attack cost.

7   Related Work

Vulnerabilities and attacks in clouds have been extensively researched. Subashini et al. [32], Hashizume et al. [19], Fernandes et al. [15], and Movsowitz et al. [29] provide extensive surveys of such attacks.

Fraudulent Resource Consumption (FRC) attacks, in which budgets are exhausted, have also been researched. Idziorek et al. [21] present an FRC attack and discuss two detection methodologies for such attacks. Kumar et al. [24] suggest an in-cloud EDoS mitigation web service (called Scrubber Service) that can be used on-demand. This service is used to generate and verify a crypto puzzle needed to prove the legitimacy for acquiring services.

Attacks on ad auctions were analyzed by Zhou et al. [38] and Cary et al. [10]. Jellinek et al. [22] study existing cloud billing systems, uncovering difficulties in predicting charges and bugs that lead to free CPU time and over-charging for storage.

8   Conclusions and Future Work

In this work, we demonstrated two low-cost economic attacks on an auction based mechanism for vertical resource allocation in the cloud. The Price Raising attack is a low cost attack that causes the victim to deplete its economic resources, thus freeing resources for the attacker to obtain at a negligible cost. We demonstrate how this attack reduces the victim's profit sevenfold. The Elbowing attack hinders the victim's performance by outbidding it every several rounds. We showed that an attacker can cause damage of \$290–\$630 for every dollar it spends on the attack. This attack can be applied to various economic mechanisms. Future work will try to amplify the effects of the Elbowing attack, e.g., by coupling it with an additional attack which will inform the attacker of optimal attack times. An optimal time for an attack like this depends on the quality and quantity of the evicted data. The RAM utilization is of high quality when the victim values its RAM usage the most. This valuation might be deduced from its bid price, or from side channels such as the victim's traffic volume or destination.

Acknowledgments. This work was partially funded by the Amnon Pazi memorial research foundation. We thank A. Schuster and E. Tromer for fruitful discussions. We thank Y. Lev, A. Ohayon, and S. Levenzon for their contribution in creating the allocation plots presented in Sect. 5. We also thank the Caesarea Rothschild Institute for Interdisciplinary Applications of Computer Science in the University of Haifa for their support.

References

1. Litke, A.G.: Memory Overcommitment Manager. https://github.com/aglitke/mom. Accessed 19 July 2018
2. Agmon, S., Agmon Ben-Yehuda, O., Schuster, A.: Preventing collusion in cloud computing auctions. In: Coppola, M., Carlini, E., D'Agostino, D., Altmann, J., Bañares, J.Á. (eds.) Economics of Grids, Clouds, Systems, and Services - 15th International Conference, GECON 2018, Proceedings. Springer, Cham (2018). https://doi.org/10.1007/978-3-030-13342-9
3. Agmon Ben-Yehuda, O., Ben-Yehuda, M., Schuster, A., Tsafrir, D.: The Resource-as-a-Service (RaaS) cloud. In: USENIX Conference on Hot Topics in Cloud Computing (HotCloud) (2012)

4. Agmon Ben-Yehuda, O., Ben-Yehuda, M., Schuster, A., Tsafrir, D.: Deconstructing Amazon EC2 spot instance pricing. ACM Trans. Econ. Comput. **1**(3), 16:1–16:20 (2013). http://doi.acm.org/10.1145/2509413.2509416

5. Agmon Ben-Yehuda, O., Ben-Yehuda, M., Schuster, A., Tsafrir, D.: The rise of RaaS: the Resource-as-a-Service cloud. Commun. ACM **57**(7), 76–84 (2014). http://doi.acm.org/10.1145/2627422

6. Agmon Ben-Yehuda, O., Posener, E., Ben-Yehuda, M., Schuster, A., Mu'alem, A.: Ginseng: market-driven memory allocation. ACM SIGPLAN Not. **49**(7), 41–52 (2014)

7. Alibaba Cloud Spot Instances. https://www.alibabacloud.com/help/doc-detail/52088.htm. Accessed 11 Mar 2018

8. Azure. https://tinyurl.com/burstable-azure-cloud-instance. Accessed 03 June 2018

9. Brandt, F., Weiß, G.: Antisocial agents and Vickrey auctions. In: Meyer, J.-J.C., Tambe, M. (eds.) ATAL 2001. LNCS, vol. 2333, pp. 335–347. Springer, Heidelberg (2002). https://doi.org/10.1007/3-540-45448-9_25

10. Cary, M., et al.: Greedy bidding strategies for keyword auctions. In: Proceedings of the 8th ACM Conference on Electronic Commerce, pp. 262–271. ACM (2007)

11. Charting CloudSigma Burst Prices. https://kkovacs.eu/cloudsigma-burst-price-chart. Accessed 21 Apr 2018

12. Clarke, E.H.: Multipart pricing of public goods. Public Choice **11**(1), 17–33 (1971)

13. Dolgikh, A., Birnbaum, Z., Chen, Y., Skormin, V.: Behavioral modeling for suspicious process detection in cloud computing environments. In: 2013 IEEE 14th International Conference on Mobile Data Management (MDM), vol. 2, pp. 177–181. IEEE (2013)

14. EC2 Instances with Burstable Performance. https://aws.amazon.com/blogs/aws/low-cost-burstable-ec2-instances/. Accessed 11 Mar 2018

15. Fernandes, D.A., Soares, L.F., Gomes, J.V., Freire, M.M., Inácio, P.R.: Security issues in cloud environments: a survey. Int. J. Inf. Secur. **13**(2), 113–170 (2014)

16. Funaro, L., Agmon Ben-Yehuda, O., Schuster, A.: Ginseng: market-driven LLC allocation. In: 2016 USENIX Annual Technical Conference, p. 295 (2016)

17. Google Cloud Platform. https://cloud.googleblog.com/2016/09/introducing-Google-Cloud.html. Accessed 11 Mar 2018

18. Groves, T.: Incentives in teams. Econ. J. Econ. Soc. 617–631 (1973)

19. Hashizume, K., Rosado, D.G., Fernández-Medina, E., Fernandez, E.B.: An analysis of security issues for cloud computing. J. Internet Serv. Appl. **4**(1), 5 (2013)

20. Hoff, C.: Cloud Computing Security: From DDoS (Distributed Denial of Service) to EDoS (Economic Denial of Sustainability). https://tinyurl.com/from-ddos-to-edos. Accessed 27 May 2018

21. Idziorek, J., Tannian, M.: Exploiting cloud utility models for profit and ruin. In: 2011 IEEE International Conference on Cloud Computing (CLOUD), pp. 33–40. IEEE (2011)

22. Jellinek, R., Zhai, Y., Ristenpart, T., Swift, M.M.: A day late and a dollar short: the case for research on cloud billing systems. In: HOTCLOUD (2014)

23. Kivity, A., Kamay, Y., Laor, D., Lublin, U., Liguori, A.: KVM: the Linux virtual machine monitor. In: Proceedings of the Linux symposium, vol. 1, pp. 225–230 (2007)

24. Kumar, M.N., Sujatha, P., Kalva, V., Nagori, R., Katukojwala, A.K., Kumar, M.: Mitigating economic denial of sustainability (edos) in cloud computing using in-cloud scrubber service. In: 2012 Fourth International Conference on Computational Intelligence and Communication Networks (CICN), pp. 535–539. IEEE (2012)

25. Lazar, A.A., Semret, N.: Design, analysis and simulation of the progressive second price auction for network bandwidth sharing. Columbia University (1998)
26. Maillé, P., Tuffin, B.: Multi-bid auctions for bandwidth allocation in communication networks. In: Proceedings IEEE INFOCOM 2004, The 23rd Annual Joint Conference of the IEEE Computer and Communications Societies, Hong Kong, China, 7–11 March 2004. IEEE (2004). https://doi.org/10.1109/INFCOM.2004.1354481
27. memcached. https://github.com/ladypine/memcached. Accessed 12 Mar 2018
28. Metz, C.: Facebook Doesn't Make As Much Money As It Could–On Purpose. https://tinyurl.com/facesbook-ads. Accessed 12 Mar 2018
29. Movsowitz, D., Agmon Ben-Yehuda, O., Schuster, A.: Attacks in the Resource-as-a-Service (RaaS) cloud context. In: Bjørner, N., Prasad, S., Parida, L. (eds.) ICDCIT 2016. LNCS, vol. 9581, pp. 10–18. Springer, Cham (2016). https://doi.org/10.1007/978-3-319-28034-9_2
30. Ristenpart, T., Tromer, E., Shacham, H., Savage, S.: Hey, you, get off of my cloud: exploring information leakage in third-party compute clouds. In: Proceedings of the 16th ACM Conference on Computer and Communications Security, pp. 199–212. ACM (2009)
31. Spot Marketing Pricing–discount Packet Bare Metal Servers. https://www.packet.net/bare-metal/deploy/spot/. Accessed 02 June 2018
32. Subashini, S., Kavitha, V.: A survey on security issues in service delivery models of cloud computing. J. Netw. Comput. Appl. **34**(1), 1–11 (2011)
33. Varadarajan, V., Kooburat, T., Farley, B., Ristenpart, T., Swift, M.M.: Resource-freeing attacks: improve your cloud performance (at your neighbor's expense). In: Proceedings of the 2012 ACM Conference on Computer and Communications Security, pp. 281–292. ACM (2012)
34. Varadarajan, V., Zhang, Y., Ristenpart, T., Swift, M.M.: A placement vulnerability study in multi-tenant public clouds. In: Jung, J., Holz, T. (eds.) 24th USENIX Security Symposium, USENIX Security 2015, Washington, D.C., USA, 12–14 August 2015, pp. 913–928. USENIX Association (2015). https://www.usenix.org/conference/usenixsecurity15/technical-sessions/presentation/varadarajan
35. Vickrey, W.: Counterspeculation, auctions, and competitive sealed tenders. J. Financ. **16**(1), 8–37 (1961)
36. Waldspurger, C.A.: Memory resource management in VMware ESX server. In: Culler, D.E., Druschel, P. (eds.) 5th Symposium on Operating System Design and Implementation (OSDI 2002), Boston, Massachusetts, USA, 9–11 December 2002. USENIX Association (2002), http://www.usenix.org/events/osdi02/tech/waldspurger.html
37. Yu, D., Mai, L., Arianfar, S., Fonseca, R., Krieger, O., Oran, D.: Towards a network marketplace in a cloud. In: HotCloud (2016)
38. Zhou, Y., Lukose, R.: Vindictive bidding in keyword auctions. In: Proceedings of the Ninth International Conference on Electronic Commerce, pp. 141–146. ACM (2007)

An Evaluation of a Market Based Resource Trading in a Multi-campus Compute Co-operative (CCC)

Md Anindya T. Prodhan$^{(\boxtimes)}$ and Andrew Grimshaw

University of Virginia, Charlottesville, VA 22903, USA
{mtp5cx,grimshaw}@virginia.edu

Abstract. Computational and data scientists at universities are often limited by the quantity and diversity of the shared resources available at their institution. Access cost for these resources are often uniform, that is, it is not differentiated based on job priority or resource requirements. This flat access policy on shared resources often lead to sub-optimal values for the institutions, and researchers with special requirements (i.e. GPU, large-memory, etc.) often have to wait significantly longer to get their job scheduled. A market-based resource trading in a multi-campus Compute Co-operative can lead to higher aggregated value for the co-operative as well as provide significant benefits for the individual institutions by scheduling jobs opportunistically when resources of one campus are over-subscribed and by placing jobs efficiently based on resource requirements. In this paper, we evaluate a resource allocation scheme in a multi-campus environment, (i.e. CCC [10]) based on job priority and resource cost, with the provision for resource trading between campuses. We collected real data traces from three (3) universities over a month and conducted a simulation to evaluate the effectiveness of our resource trading approach over the existing single institution flat rate allocation policy. Our simulation shows that, with CCC and market-based resource trading, the aggregated institutional value for the co-operative increases by 15% and the average wait time for the jobs reduce by 49%.

Keywords: Grid federation · Resource heterogeneity · Market based grid · Efficient resource trading

1 Introduction

Computational techniques, whether it is modeling and simulation or data mining, are increasingly central to research success. Research universities are facing a strategic challenge: how to provide their researchers the computational infrastructure they need to be competitive while managing costs? The problem is particularly acute because there is no single research modality. Researchers need different types of machines (tightly coupled HPC engines, high-throughput clusters, Hadoop clusters, massive data stores, GPU-based systems, etc.), at

© Springer Nature Switzerland AG 2019
M. Coppola et al. (Eds.): GECON 2018, LNCS 11113, pp. 52–65, 2019.
https://doi.org/10.1007/978-3-030-13342-9_5

different scales, and have different temporal access patterns (continuous versus bursty usage). Similarly, not all user requests are as urgent as others, and not all researchers are as well funded and able to pay for their computing. Satisfying all users' need all of the time would require an investment in infrastructure that is not possible at most institutions.

Most universities maintain a set of shared resources to support the computational needs of their researchers. Access to these resources is in most cases first come first serve (FCFS), which disregards the importance or the resource requirements of the job. Hence, "FCFS" policies on shared resources often lead to organizationally sub-optimal outcomes as not all jobs have the same value for the researchers and often the high value jobs have to wait on the queue while the low value jobs are running. As a result, funded researchers often buy their own nodes rather than using the shared resources. These private resources are often underutilized. Further, with FCFS scheduling, jobs that do not require expensive resources (GPU, largemem, etc.) are placed on these expensive, specialized, machines, in effect making jobs with those requirements wait and thus, underutilizing the full capability of those resources. Alternatively generic jobs are not executed on the specialized, expensive, resources and they remain under-utilized.

To address the problem, we have developed a market-based federation for shared resources at universities, the Campus Compute Cooperative (CCC). The CCC is a secure, standards-based, open-source, federated cloud environment that provides execution management services, data access and management services, and identity and group management services. The CCC follows the XSEDE architecture described in [5]. The CCC uses Genesis II [1,13,15] as its software stack. Genesis II is a web-services, container-based architecture that uses open standards, (primarily from the Open Grid Forum [2]) such as Basic Execution Services, JSDL, RNS, byteIO and WS-Secure Token Services.

In a nutshell, the CCC allows users to select a quality of service (i.e. urgency) for their jobs and the institutions to securely trade their existing compute and data resources. Users define their job information (resource requirements, data and executable files, command line) using a standard job description language. Once defined, jobs can be submitted to and executed on one of three different quality of service queues, high, medium, or low priority.

In the Fig. 1 users submit their job to one of three different quality of service (QoS) *gridQuues*. Each gridQueue is linked to one or more Basic Execution Services (BESes) [9]. Each BES is configured to submit jobs to a particular partition or queue in a local load management system such as SLURM, PBS, or SGE. For each BES that is linked to the gridQueue is configured with the maximum number of jobs it may concurrently submit to the BES as well as the maximum number of cores it may consume at any given instant. The gridQueue uses job resource requirements and BES resource properties to match jobs to BESes that are candidates to execute the job. The gridQueue schedules jobs to matching BESes in FCFS order.

Once a job is scheduled on a BES, the BES stages (copies) data and executables the job needs onto the selected compute resource, generates a submission

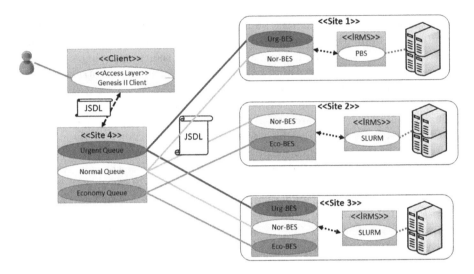

Fig. 1. In CCC, Clients send jobs to a set of gridQueues (each corresponding to a QoS specification). gridQueues then forward the jobs to appropriate BESes based on the job specification. BESes wrap compute resources such as clusters and will manage the execution of the job on the underlying clusters.

script for the local queuing system, and submits the job to the local queue. The BES then monitors the local queue until the job terminates successfully or fails. If the job completes normally the BES then stages the output files to a user specified location. The BES then notifies the gridQueue of the job status.

The effect is to give users at each participating institution access to a much larger and more diverse pool of resources than they would otherwise be able to use. This gives them access to resources their institution might not have. It also reduces average wait times significantly. Which supports the well-known queuing theory as we have replaced K independent M/M/1 queues with a single M/M/k queue.

The CCC logs all job resource and priority requests. At the end of each accounting period each institution is credited for the others jobs executed on its systems. It receives a credit based on the resources requested and the priority level at which the jobs ran. Similarly, each institution is charged for jobs executed at other institutions based on the resources requested and the quality of service (much as in Amazon EC2). For more information, including detailed documentation, downloads, and tutorials go to genesis2.virginia.edu/wiki.

In this paper we present and evaluate in simulation a market-based resource trading scheme in a multi-Campus environment like the CCC. Our hypothesis is: a market-based resource trading approach in a multi-campus environment can in fact lead to increased institutional value to the federation as a whole, while individual institutions incur significant gains. To evaluate the multi-campus resource market, we collected production data traces for a month (Oct, 2017) from three different campuses: Indiana University (IU), University of Virginia (UVA), and Virginia Tech (VT). Using the data, we conducted a simulation to calculate the *value* for each institution with and without federation.

From discussions with cluster owners at UVA we determined that one important reason the researchers buy their own nodes is so that their high priority jobs gets scheduled right away (like on-Demand access as in commercial clouds). Hence, we also measured the *average wait time* and *on-Demand capability* (number of high priority jobs that start within a threshold) for each resource.

From our results, we observe that with the CCC the aggregated value for the overall federation increases by 15%, while the average wait time reduces by 49% compared to these metrics without the federation. Also, the on-Demand capability increases from 85% to 95% with the federation, i.e. 95% of jobs placed in the high priority queue started within the 30 min.

The remainder of this paper is divided into five sections. We begin with a brief overview of the CCC, followed by a discussion of the simulation data, environment, and model. We then present and discuss the results of the simulation. Next, we compare the CCC with other market-based models, and then conclude with final observations and future work.

2   Simulation Model Inputs

To evaluate a market-based model with different resource classes and differentiated quality of service requires an input set of jobs that accurately reflects the system load we expect in production. Specifically, we would like a set of job traces that reflect user behavior in the presence of a federated environment with differentiated quality of service and diverse resources. The problem is that the system is not yet in production, so such traces do not exist. Instead we have traces from existing, non-federated environments.

A time-honored method is to generate synthetic loads or to use job traces from existing systems. The problem with purely synthetic loads is that the simulation results are only as good as your method of generating synthetic loads. The alternative of using synthetic job traces is that the job traces are collected from systems that do not have jobs that reflect new capabilities. For example, if system X has no large memory nodes then there will be no large memory job requests in the traces. Similarly, without priorities in the traced system there will be no priorities in the job traces. Our solution is to start with job traces from real systems and then transform the traces to have differentiated quality of service queues. Note that as we have not generated synthetic jobs that use non existent specialized resources we believe our simulations underestimate the value users and institutions will accrue.

We collected job traces from three institutions that are planning on participating in the CCC: UVa, VT, and IU. We collected a month of job traces from October 2017. All data was anonymized using one-way hash function on user name, group name and job name.

From UVA, we collected the job traces from *Rivanna*, the High Performance Computing (HPC) system at UVA. *Rivanna* has 240 nodes (20 cores each) equipped with infini-Band interconnect for high-performance parallel jobs. In addition to the basic nodes, *Rivanna* also has 14 nodes with GPUs (10X 4 K80s, 4X 4 P100s), and 8 with Intel Knight's Landing systems.

From IU, we collected the data traces from *Big Red II* which is the main system for high performance parallel computing at IU. *Big Red II* is comprised of 344 XE6 (CPU-only) compute nodes and 676 XK7 "GPU-accelerated" compute nodes, all connected through Cray's Gemini scalable interconnect, providing a total of 1,020 compute nodes, and 21,824 processor cores. Each XK7 node is equipped with one NVIDIA Tesla K20 GPU accelerator.

At Virginia Tech, we gathered logs from the Splunk database. The logs contain traces from four (4) distinct HPC systems available at VT: *Cascades, Dragonstooth, NewRiver* and *Blueridge*. These systems contain 817 nodes and 18,220 processor cores. Fifty-five of the nodes were equipped with Nvidia GPU, while 130 of them had 2 60 core Intel Xion Phi co-processors (mic). Four of the nodes were suitable to run large-memory jobs with a total of 3TB of memory each.

We categorized jobs into three priority groups: high priority, medium priority and low priority. Since, none of the job traces were annotated with job priorities, we decided to choose the debug jobs and jobs those were submitted to special queues (i.e. queues that are exclusive to specific research labs) as high priority jobs, which is about 5–10% of the jobs. We believe, in practice, the percentage of high priority jobs will usually be higher, which will further strengthen our claim [17]. For medium priority, we choose those jobs that require special resources (i.e. GPU, largemem, mic, KNL etc.) as they would yield more value to the institutions. The rest of the jobs are annotated as low priority. The Table 1 summarizes the traces based on their priorities for each institution.

VT logs contained 29,409 jobs with 178 large-memory jobs and over 1000 GPU jobs. IU logs contained a total of 38,115 jobs with almost a third of them being GPU jobs. IU traces did not contain any large memory job as Big Red II did not have provision for large memory jobs. With the federation, however, IU researchers can use the unused cycles of the large memory machines at UVA or VT if required. And finally, UVA logs included 66,246 jobs with 462 GPU jobs and 269 large memory jobs.

The method for categorizing jobs as high, medium, or low priority is certainly not ideal. But in the absence of data with differential quality of service there are few other options. We also examined using random assignment of jobs or a user's

Table 1. Summary of jobs based on different priorities for each institution.

Institution	High priority	Medium priority	Low priority	Total
IU	2,495	12,915	22,705	38,115
UVA	171	6,078	59,997	66,246
VT	3,149	944	25,316	29,409
Grand total	6,277	19,475	108,018	133,770

jobs to a particular queue, with form example 20% of jobs being high priority, and 30% of jobs low priority. The problem is, how do you pick the numbers?

From the traces, we found that resource utilization at UVA was 15%, which makes it the ideal candidate to lend its resource to VT and IU. However, to evaluate our approach when all the institutions have at least moderate load, we evaluated the system with and without UVA.

3 Simulation Model

Using the job traces we simulate a federated queue of queues priority scheduler in which jobs are placed into one of three global queues, high priority, normal priority, or low priority. The resource pool consists of all the resources from different institutions. Each resource is annotated with a set of resource properties which specifies the special features of the resource (GPU, largemem, KNL, inifi-Band), and the institution it belongs to. The per unit cost of a resource is determined based on these properties.

When a new job arrives in the queue, the scheduler calculates a candidate set of resources that the job can run on and sorts the list based on resource cost. To schedule a job, the scheduler picks a job from the top of the highest available priority queue and goes through the candidate set to find a resource that can run the job at that time. If none of the resources from the candidate list can run the job at that time, the job is put back to the queue and the scheduler picks the next job from the queue.

The actual CCC queues [10] are configured with a set of BESes (cite) resources on which they each can place a defined maximum number of jobs and use a maximum number of cores. The BESes each wrap a particular local queue on a particular resource. Thus a machine such as Rivanna at UVA can be configured to have several BESes: a high-priority single node job queue, a high-priority multinode (MPI) queue, a high-priority GPU queue, a high-priority large memory queue, a low-priority single node job queue, a low-priority MPI queue, and so on.

3.1 Model Definitions

Resource Pricing: For resource pricing, we use a basic node configuration and assign a base price to the configuration. Then, we put additional charges for each additional feature, e.g. GPU, added to the base configuration. Resource pricing is a major factor in job value, so prices should reflect prices in the real world. We follow the Amazon AWS on-Demand pricing scheme.

Table 2 summarizes the reference instance chosen from AWS, its price and ratio and our multiplier chosen from the ratio. For our experiment, we chose a multiplication factor of 5 for GPU resources as in the CCC the GPU resources have more cores per machine. The large memory machines in the CCC have a factor of 3. Since there was no reference to valuate infini-band interconnect, we chose a factor of 2 for that. For special features like mic or KNL, we chose a

Table 2. Resource pricing based on Amazon AWS on-Demand Pricing

Function	Ref instance	Cores	Memory	GPUs	Price/core-hr	Ratio	Multiplier
General purpose	m5.4xlarge	16	64	0	0.0480	-	1
GPU	p2.xlarge	4	61	1	0.2250	4.68	5
Large memory	x1.32xlarge	128	1952	0	0.1042	2.17	3

multiplication factor of 3, as they are not as powerful as a GPU, however they provided extra co-processors for faster execution.

Job Price: Job price is defined as the price that the user must to pay[1] for executing a job on the market resources. The price of a job depends on three factors: (i) requested resource type, (ii) execution time on the resources and (iii) QoS requested by the user. Formally,

$$job\_price = req\_res\_cost \times exec\_time \times priority\_factor \qquad (1)$$

For the experiment, we chose a *priority_factor* of 4 for high priority jobs, and 1 for the rest. Medium priority jobs will have higher value than the low priority jobs because of their special resource requirements.

Job Value: Job value is defined as the value the user gets by running a job. Job price is a lower bound on the value that the researcher gets as the researcher is at least willing to pay *job price* for running the job. (The surplus value could be higher if the PI would have been willing to pay more).

Institutional Value(ins_val): Institutional value is the sum of the values of all the jobs run through the shared resources. Formally,

$$ins\_val = \sum_{all\_jobs} job\_value \geq \sum_{all\_jobs} job\_price. \qquad (2)$$

However, we believe the job will have value to the researcher only if job is finished before some deadline. So, we have decided to annotate each job with a deadline based on the wall-time requested for the job. For high priority jobs however, the job will only have value if it starts immediately (within a threshold). For our experiment we choose the threshold as 30 min. Equation 3 defines the deadline of the ith job.

$$deadline_{job_i} = \begin{cases} 30 * 60 + exec\_time_{job_i}, & \text{if } job_i \text{ is a high priority job} \\ 2 * walltime_{job_i}, & \text{otherwise} \end{cases} \qquad (3)$$

Hence the *job_value* and *ins_val* can be formally presented by Eqs. 4 and 5 respectively

$$job\_value_{job_i} = \begin{cases} job\_price, & \text{if } finish\_time_{job_i} \leq deadline_{job_i} \\ 0, & \text{otherwise} \end{cases} \qquad (4)$$

---

[1] Whether users pay or not is up to their institution. The users' institution is responsible for paying in kind or paying with cash.

$$ins\_val = \sum_{job_i \in all\_jobs} job\_value_{job_i} \tag{5}$$

In order to have a better realization of ins_val, we decided to represent ins_val into dollars. Once again, we used AWS pricing as reference to convert ins_val into dollar value. Since the AWS does not provide the QoS breakdown as CCC, and AWS on-Demand execution would be comparable to a high priority execution of a job, we chose the value for a high priority job with no special features to be 4.8 cents-per-core-hr. So, the value for a general-purpose medium or low priority job would be 1.2 cents-per-core-hr. One thing to note here is that, we are ignoring the storage cost and data transfer cost of Amazon AWS for our calculations. Hence, we are actually under-valuing the jobs slightly. From here on we will use the dollar value as the institutional value for clarity.

$wait\_time$: The $wait\_time$ of a job can be defined as the difference between the $queue\_time$, that is the time when the job was placed in the queue and the $start\_time$, that is the time when the job started execution. So, the average wait time can be expressed by the following Eq. 6

$$avg\_wait\_time = \frac{\sum_{job_i \in all\_jobs}(start\_time_{job_i} - queue\_time_{job_i})}{num\_of\_jobs} \tag{6}$$

The percentage of high priority jobs scheduled within the defined threshold (%high) is used to evaluate the on-demand access facility for the high priority jobs and can be expressed with the Eq. 7.

$$\%high = \frac{\text{number of high priority jobs finished within deadline}}{\text{total number of high priority jobs}} \times 100\% \tag{7}$$

4   Results and Analysis

We used the Gridsim [8] based job scheduling simulator ALEA v4.0 [12] to simulate job scheduling. We first calculated a baseline institutional value for each institution by "running" their jobs on their machines only (without federation). We also measured the baseline values for avg_wait_time, and %high. Then we calculated these metrics with federation and compared the results.

Figure 2 presents the institutional value, average wait time and percentage of high priority jobs started within the threshold respectively for each of the three institutions individually and combined (the last set of bars in each figure). Since the job traces from UVA had only about 15% resource utilization, we wanted to evaluate the federation excluding UVA, to understand whether institutions can gain anything even if all the resources are highly utilized.

From Fig. 2(a), we can observe that, both IU and VT gains institutional value from the federation. IU gains about 2% from the federation excluding UVA and about 12.5% with the full federation. VT on the other hand gains significantly more from the federation. VT gains 14% from the federation when

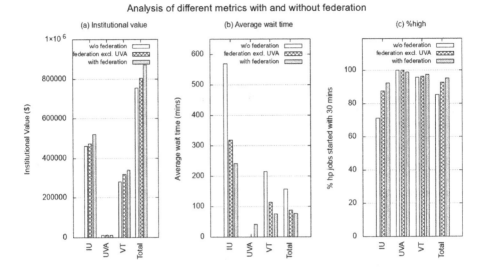

Fig. 2. With a multi-campus market based resource trading scheme in CCC, (a) all the institutions in the federation can gain values by trading of their resources when they are under-utilized or by getting the jobs completed before deadline, (b) the wait time reduces significantly with the federation with the wider and more diverse set of resources, and (c) more high priority jobs are scheduled within the predefined threshold

UVA is excluded, while 21.5% with the full federation. Finally, the overall value of the federation increases by 6.6% with a federation between IU and VT and 15.6% with the full federation. In all the cases VT seems to be the institution with most benefits from the federation. This is because, VT resources are often over-subscribed (overall utilization is >92%) and as a result many of the jobs miss their deadline waiting on the queue when their resources are occupied. A major portion of these jobs are serial or MPI jobs which can very easily be scheduled at UVA or IU. For IU data traces, we observed that many of GPU jobs were missing their deadlines because of the large backlog in the GPU resources.

Since GPU resources are scarce in all other institutions the benefit for IU was limited when UVA was excluded as VT GPU resources are most often than not occupied. However, when UVA was included in the federation, because of its very low utilization on the GPU resources, we could see a much more significant gain for IU. On the contrary, the institutional value of UVA goes down by about 1% as jobs from VT and IU forces some of the UVA jobs to miss their deadlines. Under this condition, one might wonder, what is the motivation for UVA (or IU in the case of federation excluding UVA) to participate? The motivation is the ability to trade their unused resources to the other participants of the federation (i.e. VT) for money or in-kind resource use of resources they do not have.

The accounting and clearance mechanism facilitates adjusted compensation to the resource providers (The net between the cost of resources they provided others and the cost of resources they used of others). Table 3 shows a summary of trade between institutions. From the table we can observe that for the month, UVA is owed \$31,658.52 and \$16,832.01 from the VT and IU respectively and IU is owed \$8,318.76 by VT. VT gains a value worth of 5,028,026.26 core-hrs from the federation, which would have cost about \$241k from Amazon AWS, at a cost of \$40k. Further, with federation, VT gets this value at about 70% utilization of their resource. So, this provides opportunity for VT researchers to submit additional jobs which would not have been completed without the federation or VT can rent these unused portions to the other institutions if more institutions join the federation. On the other end of the spectrum, UVA will earn more than \$48k by lending their unused resources to the federation in a month.

Table 3. A summary of the accounting and clearance results with the federation

Job originator	Executed at	Trade amount	Surplus	Balance
IU	UVA	\$18,170.03	UVA	\$16,832.01
UVA	IU	\$1,338.03		
IU	VT	\$59,079.25	IU	\$8,318.76
VT	IU	\$67,398.02		
UVA	VT	\$2,731.03	UVA	\$31,658.52
VT	UVA	\$34,389.54		

Figure 2(b) illustrates the average wait time for the jobs at different organization both in the case of without federation and with federation. As expected, the wait time for jobs at VT and IU are reduced significantly via federation. For VT jobs, the average wait time reduces by 47% through a federation with IU and by 65% with the full federation. The average wait time for the IU jobs were reduced by 44% with the federation excluding UVA and 58% for the full federation. Overall the federation reduces the average wait time for all the jobs by 49%. However, for UVA job traces the average wait time increases from 6 s to about 41 min. The reason behind is that the UVA resources are currently under-subscribed, which means without federation most of the jobs can start right away. With the federation, since UVA would be trading its unused resources to VT and IU, the wait time for UVA jobs increases, specially the GPU jobs suffer most as there are huge demand for GPUs at both IU and VT and UVA itself has only 14 GPU nodes.

Another metric is the percentage of high priority jobs (%high) scheduled before a predefined deadline. Figure 2(c) presents this percentage for different institutions. From the figure we can note that for both VT and IU %high increases with the federation, while for UVA the percentage reduces from 100% to 98%. The reason for the decrease in %high for UVA is that VT and IU will fill UVA resources, and as UVA includes many small jobs which can't be moved around, some high priority jobs miss their deadline.

4.1 Data Transfer Overhead

A potential problem with the current simulation results is that the simulation ignores the network overhead due to the transfer of data from one site to another. This is because the job traces did not include the input and output data sizes that needs to transferred with the job. In future with CCC, we are planning to instrument the amount of data needed to move for each job to model the data transfer. However, in **citetechReport**, Vanamala et al. reports that it is possible to transfer a gigabyte of data in 2 to 4 s in a wide are network. Hence, if the data transfer requirements are not in the TB range, data transfer overhead will contribute only a few minutes at either end of the job. As the average wait times for the job traces are more two and half hours in any case, we believe adding a few minutes upfront (and at the end) will not significantly change the results as CCC cuts the wait time in half. Furthermore, the job submission language, JSDL, supports the notion of local scratch spaces where users may cache large input files that may be reused across many runs. In other words, if ten jobs to be run at a site each need the same 100 GB file it only needs to be transferred once to the site. Which further reduces the average data movement overhead.

5 Related Work

There is a vast related literature in both computational grids and grid economic models going back over 40 year. Here we will focus on contemporary cloud and on-demand computing, grid economic models, and an well known grid federation alternative, Open science grid (OSG).

Since its emergence cloud computing has been more successful in the commercial space compared to Grid computing or earlier metasystems efforts. The main reasons for this unprecedented success include but not limited to, (i) the use of virtual machines as sand-boxes which gave users tremendous flexibility; (ii) the rise of pay per use on line media (movies, music), social networking, and other areas that paved the way for market acceptance; and (iii) the entrance of Amazon into the market place, a vendor that was not only trusted but assumed to know what it was doing.

NIST defines clouds as having five essential characteristics [14]: on-demand self-service, broad network access, resource pooling, rapid elasticity, and measured service. The CCC fits under the National Institute of Standards and Technology (NIST) definition of a cloud. NIST further defines three service models, Software as a Service (SaaS), Platform as a Service (PaaS), and Infrastructure as a Service (IaaS). The CCC straddles the definition of a PaaS and IaaS cloud. The CCC allows users to deploy applications that may or may not depend on CCC services, and in the near future will include the ability to deploy and execute virtual machines.

There are several vendors of cloud services. The two most prominent are Amazon Elastic Computer Cloud (EC2) and S3 services and Microsoft with the windows Azure Cloud. The CCC differs from these commercial cloud providers in two significant ways. First, in the CCC economic actors are both producers and consumers. When using commercial clouds, it is clear that the cloud provider is selling and the user is buying. Thus, when interacting with commercial cloud providers there will always be a bill to be paid. In the CCC, bills can be paid in-kind through resource trading. Second, new resource providers can join the community (enlarging the resource pool) without any change required on the client side.

The literature on grid economic models is huge, for example [7,9,11,20,21]. The CCC builds on and extends these earlier systems. The CCC is based on open standards developed in the Open Grid Forum [2]. One of the defining characteristics of most early metasystems and grid efforts was that resources were "free" or that users already had to have permission to use the resources.

In terms of grid economics [4,7,16], most of the work has been done in simulation, or developed but not used in a production environment. Several research systems like: Spawn [19], AppLes [6], Popcorn [18], G-commerce [21], OCEAN [21], Nimrod [3], GridSim [8], and GridEcon [4] have explored the use of different economic models for managing resources in grid environment. In each of these systems a market is established to trade resources between the resource providers and resource users. But none of these actually considered the quality of service requirements of the users. Buyya et al. [7] proposed a QoS based scheduling algorithm but only considered the deadlines as a QoS measure, but did not consider the value gained from resource diversity.

The CCC is different in that our explicit goal is to construct and operate a production quality market where the actors buy and sell computational services to test the hypothesis that markets combined with differentiated quality of service result in a win-win for all actors. Hence our work is based on production level data traces on real production clusters.

Scientists have many different options for deploying their applications, including grids such as Open Science Grid (OSG) or XSEDE, or academic clouds like the FutureGrid and Red Cloud, and many super-computing centers. The basic problem with these resources is the fact that these resources are mostly used as a best effort basis without regarding the user QoS requirements or resource costs, which leads us back to our original problem.

6 Conclusion

In this paper we presented the simulation results of a small cross campus computing environment consisting of three universities. The simulations focused on *institutional value, average wait time*, and *the number of high priority jobs that started within 30 min.* with and without resource federation. The objective was to support our claim that, "multi-campus market based resource trading with CCC can provide better value for each of the institutions with their existing set of shared resources".

The results clearly show that federation, i.e. a single global queue, increases institutional value, decreases wait times, and increases the number of high priority jobs that start within 30 min. Insofar as universities have idle cycles (as UVa does), these cycles can be consumed with significant benefit in the form of additional capacity at all institutions. In particular note that VT load has decreased from 92% to 70%, representing freeing up of 2.8M core hours of capacity that can be used for more applications.

There are limitations to these results. Primarily, the results do not show the benefits of differentiated quality of service, which we believe is substantial. This is because the traces do not include all of the attributes, such as priority (for differentiated quality of service), that we need to produce accurate simulations. In particular, we believe that by underestimating the number and duration of high-priority jobs. Similarly, the site traces do not include jobs that use resources not present at the site, underestimating the diversity of jobs we expect when federation is complete and visible to end users.

Our future work will focus on these short-comings, specifically coming up with a realistic and defensible method to generate job priorities and job diversity for traces where they do not exist. With these enhanced traces we will be better able to predict institutional gains, in particular institutional gains for institutions that have not yet joined the cooperative.

Acknowledgments. We would like to thank several people, from Indiana University Craig Stewart and Rich Knepper for leading the Campus Bridging effort, and Matt Allen for gathering the Big Red job files. We would also like to Mark Gardner from Virginia Tech for his continuous support on the project and gathering the VT traces for us. This document was developed with support from National Science Foundation (NSF) grant OCI-1053575. Any opinions, findings, and conclusions or recommendations expressed in this material are those of the authors and do not necessarily reflect the views of the NSF.

References

1. The genesis ii project, virginia center for grid research. http://genesis2.virginia.edu/wiki/. Accessed May 2018
2. Open grid forum. https://www.ogf.org/. Accessed May 2018
3. Abramson, D., Sosic, R., Giddy, J., Hall, B.: Nimrod: a tool for performing parametrised simulations using distributed workstations. In: Proceedings of the Fourth IEEE International Symposium on HPDC, pp. 112–121. IEEE (1995)
4. Altmann, J., et al.: GridEcon: a market place for computing resources. In: Altmann, J., Neumann, D., Fahringer, T. (eds.) GECON 2008. LNCS, vol. 5206, pp. 185–196. Springer, Heidelberg (2008). https://doi.org/10.1007/978-3-540-85485-2_15
5. Bachmann, F., Foster, I., Grimshaw, A., Lifka, D., Riedel, M., Tuecke, S.: XSEDE architecture level 3 decomposition. Version 0.972, June 2013
6. Berman, F., Wolski, R., Figueira, S., Schopf, J., Shao, G.: Application-level scheduling on distributed heterogeneous networks. In: Proceedings of the 1996 ACM/IEEE Conference on Supercomputing, pp. 39–39. IEEE (1996)
7. Buyya, R., Abramson, D., Venugopal, S.: The grid economy. Proc. IEEE **93**(3), 698–714 (2005)
8. Buyya, R., Murshed, M.: GridSim: a toolkit for the modeling and simulation of distributed resource management and scheduling for grid computing. Concurr. Comput. Pract. Exp. **14**(13–15), 1175–1220 (2002)

9. Grimshaw, A., et al.: An open grid services architecture primer. Computer **2**, 27–34 (2009)
10. Grimshaw, A., Prodhan, M.A., Thomas, A., Stewart, C., Knepper, R.: Campus compute co-operative (CCC): a service oriented cloud federation. In: 2016 IEEE 12th International Conference on e-Science (e-Science), pp. 1–10. IEEE (2016)
11. Grimshaw, A.S., Wulf, W.A., et al.: The legion vision of a worldwide virtual computer. Commun. ACM **40**(1), 39–45 (1997)
12. Klusáček, D., Rudová, H.: Alea 2 - job scheduling simulator. In: Proceedings of the 3rd International ICST Conference on Simulation Tools and Techniques (SIMU-Tools 2010). ICST (2010)
13. Koeritz, C.: Genesis ii omnibus reference manual. Technical report, Genesis II Group, University of Virginia (2012)
14. Mell, P., Grance, T.: The nist definition of cloud computing. Commun. ACM **53**(6), 50 (2010)
15. Morgan, M.M., Grimshaw, A.S.: Genesis ii-standards based grid computing. In: Seventh IEEE International Symposium on Cluster Computing and the Grid, CCGRID 2007, pp. 611–618. IEEE (2007)
16. Padala, P., Harrison, C., Pelfort, N., Jansen, E., Frank, M.P., Chokkareddy, C.: Ocean: the open computation exchange and arbitration network, a market approach to meta computing. In: International Symposium on Parallel and Distributed Computing, p. 185. IEEE (2003)
17. Prodhan, M.A., Grimshaw, A.: Market-based on demand scheduling (MBoDS) in co-operative grid environment. In: Proceedings of the 2015 XSEDE Conference, p. 26. ACM (2015)
18. Regev, O., Nisan, N.: The popcorn market. Online markets for computational resources. Decis. Support. Syst. **28**(1), 177–189 (2000)
19. Waldspurger, C.A., Hogg, T., Huberman, B.A., Kephart, J.O., Storn, W.S.: Spawn: a distributed computational economy. IEEE Trans. Softw. Eng. **18**(2), 103–117 (1992)
20. Weitzel, D., Bockelman, B., Fraser, D., Pordes, R., Swanson, D.: Enabling campus grids with open science grid technology. In: Journal of Physics: Conference Series, vol. 331, p. 062025. IOP Publishing (2011)
21. Wolski, R., Plank, J.S., Bryan, T., Brevik, J.: G-commerce: market formulations controlling resource allocation on the computational grid. In: Proceedings of 15th International Symposium on Parallel and Distributed Processing, 8-pp. IEEE (2001)

Snooping Around a Fence:
A Lesson from the Education Sector
in a Software Service Ecosystem

Djamshid Sultanov[1(✉)], Kibae Kim[2], and Jörn Altmann[1]

[1] Seoul National University, Seoul, South Korea
sdjamshid@gmail.com, jorn.altmann@acm.org
[2] Korea Advanced Institute of Science and Technology, Daejeon, South Korea
kibaejjang@gmail.com

Abstract. Although the education sector has recognized the value of information technologies since the early 1990s, the advancement of education services is not clearly shown in the information technology era. This paper visualizes the trace of education services development in a software service ecosystem with real data about software services and their combinations resulting in composite services. Our graphical analysis results show that education services continuously emerge through reusing and recombining popular software services such as Google Maps and Facebook, although only a few education software services open their functions and data to the ecosystem. Moreover, our analysis results show that there no service groups that are built around education services. Our findings suggest that the education sector is immature within the software service ecosystem and that a software service sector has not been formed yet.

Keywords: Education · IT adoption · Software service · Network analysis

1 Introduction

Education has been adopting information technologies (IT) since the early 1990s [1, 2]. IT technologies support the access to information, simplifies the knowledge transfer at classrooms [3], and enables equal opportunities for students in remote areas and for disabled students [4]. This service is not just expected to be provided to secondary and tertiary students but also students at early childhood [5].

Although the prior work suggests various opportunities of IT adoption in education, a few recent studies raise issues on the market response to IT in the education sectors. The continuation of the IT transformation of classrooms is doubtful, if the government adopts IT due to technological possibilities instead of market demand [6]. The emergence of breakthrough innovation is also in question, if the market demand does not pull the technologies [7, 8].

Our motivation of research is to address the issue of market response to IT adoption in education. The market demand drives the evolution of technologies by recombining previous technologies and new technologies [8]. Thereby, we conjecture that the

© Springer Nature Switzerland AG 2019
M. Coppola et al. (Eds.): GECON 2018, LNCS 11113, pp. 66–76, 2019.
https://doi.org/10.1007/978-3-030-13342-9_6

evolutionary path reflects the market demand on the synthesis between information technologies and education services, and the extension of the synthesis to another technology fields (e.g., finance, manufacturing, and transportation).

In this research, we address this by investigating the evolution of education software services in a software service ecosystem. Several recent studies suggest that software services advance through interacting with each other in an entire ecosystem. They diagnose services' creativity and sustainability and forecast the future facets of technologies and societies [9, 10]. In a similar line, we address how much an education sector embedded in the software service ecosystem, and what technologies the education services mainly interact with.

To investigate this, we aggregate empirical data from www.programmableweb.com, which lists the information of software services (represented as APIs) and their use in composite services (represented as mashups). Our data set consists of 127 composite services in education sector and 421 software services that are used to develop those composite services between 2006 and 2017. In a graphical view, we define a software service ecosystem as a set of vertices, representing software services, and their edges, indicating co-reuse of software services in a composite service. We measure the network position of an education service with centralities, indicators of social network analysis, and their memberships in clustered subgroups.

Our analysis results show the followings. First, new education services emerge through the convergence of software services belonging to other sectors than the education sector. Second, only a few software services belonging to the education sector enter the software service ecosystem. Rather, education services are created around a few popular software services such as Google Maps, Facebook, Flickr, and Twitter. Third, the education sector looks declining in recent years. On the ground of those findings, we carefully conclude education is snooping around a fence to IT adoption but did not yet hurdle over the fence.

Our findings contribute to both academia and business. From an academic perspective, our findings suggest a graphical analysis of well-known social network analysis tools reveals the actual shape of the IT adoption of education. Our findings show the real innovation could be different from expectations that are based on theory. Education does not actually reap to the core of the software ecosystem, while the IT adoption is believed to be universal in a long history and provide opportunities [1–5]. This academic contribution leads to a managerial implication. The innovation through the convergence between education and another technology should be carefully designed on the ground of the market demand, while the academic research underlines the opportunities of new technologies.

2 Theoretical Background

2.1 IT Adoption in Education

Education is the communication on a specific subject between a teacher and students and among students [11, 12]. A talk in a certain physical place is a typical way of communication in the Ancient Greek schools, the medieval universities, and in a modern classroom [13]. As books remove the temporal and special constraints of communication, information of a teacher could spread quickly and widely to any students, who can buy a book and read it [12]. Now, education practitioners and experts in recent two decades paid attention to information technologies (IT) that support the communication among people, which is just the key to education, and fast adopt the IT in education for distant communication, easy modification of teaching materials, and visualization of its object [1, 14–18, 54].

The IT adoption in education transforms the way of teaching and learning within classrooms as well as out of classrooms. First, the IT mediates the communication between a teacher and students, and among students in a classroom [1, 2, 15, 16, 19–21]. If the traditional education focuses on unidirectional knowledge flow from a teacher to students, the IT in a classroom extend the education into constructivist, socio-cultural, cognitive and collaborative ways [2]. In detail, personal computers connected to each other through Internet support the students share their knowledge to build new knowledge by their practice [19, 20]. Multimedia-systems and simulation environment, as well as, help students' cognitive experience in concrete objects that they learn [2, 17, 22]. Those promote provides students with creative thinking, while the conventional one stops at just bringing knowledge to students [5].

Second, the IT reduces the spatial and temporal restrictions in communication between a teacher and students [2, 17, 19, 20, 22–30]. Through the Internet, the education service can also be provisioned to students uncomfortable to move [4], and living in a distant rural area such as Outback of Australia [17]. The video conference system and online transmission of class materials makes it indistinguishable between a physical classroom and a screen [22]. On the ground of Web services, furthermore, a teacher provides their service to a bigger market, which has no size limitation, than a classroom of 50 students at most [15]. As the lecture is stored in a Web server, a student, who hard to match their schedule to class, can attend the class on screen at their convenient time [31].

In summary, the IT adoption in education removes the physical, spatial and temporal barriers to the interactive and distant communication. This technological advance attracts the IT entrepreneurs to provide facilities that support teaching and learning [17], and extends the beneficiary group of education [32–34], although it does not provide all functions of traditional education such as the emotional interaction through physical touch [23]. The economy of scale in education then reduces the price of education [22], and potentially increases the quality of the education service [18]. If the IT opens a new market in education, the remaining issues are now whether and how actively the market pull the innovation [8].

2.2 Software Service Ecosystem

A software service is software that is provisioned as a service to support the interaction between a person and a machine, and/or among people on the Web [9, 10]. For example, end users access the server of Google Maps that contains map data and related functions through a Web browser to read a map of Waco in Texas. These end users do not need to install a standalone package with a map on their personal devices [35]. Amazon provides even computation as a service, which was previously provided as a product (i.e. a personal computer and a shared server); a user does not need to know the location of computers and pays for the service on demand [36].

An advanced way of using software services is accessing to the data repositories and computation resurces through a standardized interface, or generally called an open application programming interface (open API) [37]. A user can then automate using the functions that a software service provides, so that they embed the functions in its own service as its components. The access to the software services through the open API leads promoting innovation in an open manner. That is, a third party user creates a new service, which is called a composite service, by adding their own data and functions on top of one or more software services shared through the Internet [10, 37]. In this way, various servcies in a "long tail" can be released, with reducing the burden on a huge amount of investment in basic functions such as map data, search engine, data storage and servers for computation [38].

The new style of innovation builds an ecosystem of software services. Software services support creating composite services; composite services satisfies end users' demand, and composite services feed the economic return to the software services they are based on [10, 39]. Furthermore, some software services supplement another software services, as well as compete with them. Although Yahoo is a competitor of Google in the search engine market, for example, search engines of the two rivals are both used in creating Maps Compare, in which one covers what the counterpart does not take [40]. Kim et al. [10] named it a "software service ecosystem", the intertwined relationship among software services and composite services, on the analogy to the ecosystem of animals, plants and fungi that form a complicate set of competitive and symbiotic relationships.

Education is one of the sectors of software services that consist of the software service ecosystem. By December 2017, 421 software services were released to cover education among around 18,000 software services in all sectors, and 127 composite services were developed with software services during the same period [40]. Those services extend the area of IT in education from communication within and over a classroom to anything related with education, including knowledge management and education administration. For example, Mendeley opens its functions third parties for supporting scholars to manage their literatures and collaborate in writing an article [41]. UC Berkeley opens its data for education services (e.g. applicants' status, class information, and so on) through open APIs [42]. A remaining issue is then how vigorous the innovation is through the participation of software service providers and third party developers.

2.3   Diffusion of Software Services

The market needs time in adopting a technology, but a technology has a limited longevity in the market. A technology shows a bell shape curve from its birth to death through prosperity as inventors and imitators in a limited population adopts the technology [43, 44]. Although an old technology fades out in the market at the end of its longevity, a successive technology replaces the old one to continue the growth of industry as long as the market demands it [45]. However, all technologies do not surf on this life cycle successfully. Even an advanced technology can fail to attract the majority of consumers, if it does not satisfy the market demand in front of the "chasm" between early adopters, who responds to technological opportunities, and the early majority, who fulfill their own demand [46].

Some software services show the bell shape curve of lifecycle of connectivity in the software service ecosystem [10]. The market responds to a software service in two ways. End users directly use a software service through the Web pages on their demand. As well as, third party developers reuse the software service to create composite services that satisfy end users. In this line, we can make an analogy of the relationship between software services and composite services to the relationship between invention and imitation of technologies. Remaining issues are then how actively the market responds to the education software services, whether they surf on the lifecycle like successful software services or snoop around the chasm to decline at last.

3   Methodology

Our data of software services and composite services are aggregated from http://www.programmableweb.com [9, 10]. This website provides the information of software services that open their APIs and mashups that use those software services with open APIs. Around 18,600 software services and 7,900 composite services are listed in the website and sorted into 482 service sectors. Among the software services released between September 2005 and December 2017, we select 421 software services and 127 composite services that belong to the education service sector.

We apply a social network approach to those empirical data of software services and composite services. Each software service is represented with a vertex in a network graph. We consider an edge is formed between a pair of vertices if software services corresponding to those vertices are used together for developing a composite service. Each vertex contains its attribute information of the provider (e.g. Google, Yahoo, and Amazon), the service sector (e.g. mapping, social networking, education) and the release date of the corresponding software service. Each edge has no information of direction, and contains the weight meaning the number of concurrent use of the corresponding pair of software services for developing composite services.

We measure three indicators in the software service network to determine whether the software services in education are just snooping around a fence of chasm or hurdling over the fence. The first indicator is the number of software services that newly enters the market for each month. Releasing a software service requires the data and function to be shared, and the motivation of the service provider to share its service functions [47]. For example, UC Berkeley opens its API to the public because it has systematically accumulated the data of its education experience, and its sharing strategies potentially attract more and better scholars thanks to its enhanced utility of the university members through the convenient education services [42]. Therefore, the number of software service in education indicates the technological maturity that promotes the innovation in education through responding the fine demand of the market.

The second indicator is the position of a software service that is measured on the ground of the edges in the software service network [10]. We apply two indicators of social network analysis, or degree centrality and betweenness centrality to the software service network of education [9]. Degree centrality of a vertex is the number of edges that are attached to the vertex. This means how frequently a software service is used together with another software services for developing composite services. Betweenness centrality of a vertex is the number of shortest paths of a pair of vertex that passes by the vertex divided by the number of all shortest paths connecting the pairs of vertices. This implies how much the software service connects the entire parts of the software service network of education.

The last indicator is the existence of clusters of software services that rally around education service sectors. A cluster in a network is a group of vertices that are connected with each other more densely than with the vertices out of the group [48]. A cluster means that vertices belonging to it share some properties such as same opinions in case of a social network [49]. Likewise, a cluster in a software service network represents a latent sector that software services contribute to complementarily together. We implement the leading eigenvector algorithm to detect clusters in the software service network of education [50].

4 Analysis Results

Figure 1 describes the annual trend of the number of software services and composite services between 2006 and 2017. The number of software services released in education sector soared up from 11 in 2010 to 93 in 2012 and decreased afterwards. On the other hand, around 20 composite services were developed in the education sector between 2006 and 2010, and the annual number of composite services dropped from 29 in 2010 to 6 in 2011 with remaining stable afterwards. The results suggest that there was a boom of releasing software services around 2012, but the innovation on the ground of software services does not follow their release. This boom in education is not so small comparing to the number of software services in the entire sectors. During the study period, 421 software services were released in the education sector, while each service sector contains 38.6 software services (i.e. 18,600 software services are released in 482 service sectors during the same period).

Table 1 depicts six representative software services. The first five software services are most frequently used for developing composite services in education (Google Maps, Facebook, Twitter, Flickr, and YouTube). Google Maps is the software service at the top of reused software services among all sectors as well as in the education sector. This result is consistent to the description of Kim et al. [10], which shows the software service ecosystem evolves mainly on the ground of Google Maps as a platform combined with photo and video services in the early periods, and social networking services in the later periods. DonorsChoose, an online charity that promotes students who needs support [40], is the most frequently used software service among education software services. Those results suggest that innovation in education services do not frequently reuse software services in education sector. Instead, the innovation is mainly led by the software services that are most frequently used in the entire software service ecosystem.

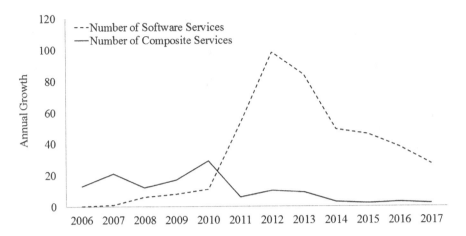

Fig. 1. Trend of the number of software services and composite services

Table 1. Top 7 reused software services over time

Service name	Release date	Provider	Service sector	Number of reuses in all sectors	Number of Reuses in the Education Sector
Google Maps	December 2005	Google	Mapping	2578	66
Flickr	September 2005	Yahoo	Photos	635	11
YouTube	February 2006	Google	Video	707	9
Facebook	August 2006	Facebook	Social	451	9
Twitter	December 2006	Twitter	Social	826	7
DonorsChoose	March 2009	DonorsChoose	Education	16	14

Figure 2 depicts the position of software services in the map spanned with degree centrality and betweenness centrality. Google Maps is connected with most of software services in the entire system as well as mediates the connection of a majority of software services. Facebook follows the position of Google Maps according to degree centrality and betweenness centrality. We call the vertex at the position that mediates the entire network with rich connectivity a "hub" [9, 51]. The entire network is rigid as long as the hub works without errors [52]. Although the hubs in the entire ecosystem play the role of hubs in the education sector, we do not see any education software services at a central position near to the hubs. DonorsChoose is the most frequently used software service in the education sector, but it is located far from the hub position on the map of education software ecosystem. In other words, the software service ecosystem of education maintains mainly through the reuse of and recombination with software services out of the education sector.

Figure 3 shows the network of software services connected through concurrent reuse for developing composite services. The color of vertices represents the membership in clusters, and their sizes are proportional to their frequency of reuse. Five clusters are detected in the main components, and three clusters in the small independent components. The six representative software services belong to each of the clusters in the main component which are distinguished by different grey-scale. However, no education software services contribute to the connectivity of each cluster in the main component. Only a small independent component consists of education software services: Finalsite, Schoology and Whiplehill, and the cluster is formed by developing one composite service (CustomSync for The Education Edge).

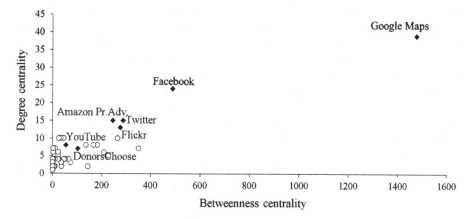

Fig. 2. Betweenness and degree centrality map

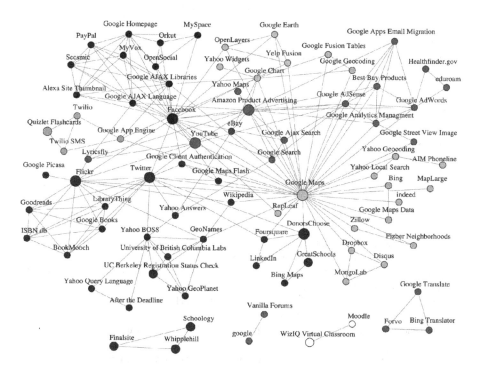

Fig. 3. Co-reuse network map of software services (All periods 2006–2017)

5   Concluding Remarks

Our analysis results show that the innovation of education services is not so impressive as could be expected according to the opportunities that IT shows in education [1, 2]. We expected that an education sector is much likely to be built in the software service ecosystem, because the education sector has adopted a variety of information technologies since 1990s [1, 2, 15], and any society must underpin the education to maintain its sustainable growth and support the competence of individuals. That is, technologies are introduced to the market; i.e., an amount of software services opened their APIs. And the market expectedly demands them. However, our analysis results show that creating composite services do not rely on education software services but on software services in other sectors.

Our findings suggest education software services do not hurdle over but still snoop around the fence to the diffusion of innovation [46]. Our findings require explanation because they are bizarre in the sense of prior theory both to IT adoption in education and the diffusion of innovation [46, 53]. In our further studies, we will discuss the reasons of snooping-around-a-fence from three perspectives. First, the education sector is waiting for killer services like Facebook and Twitter did in the social networking service sector [10]. Second, software services might originally be interdisciplinary. Therefore, an education service is much likely to be invented by combining one or more software services in other sectors. Finally, the IT adoption in education might be driven by a political need instead of market demand, so that the education sector does not emerge in the software service ecosystem, where the market demand dominates.

Acknowledgement. This research was supported by the Basic Science Program of the National Research Foundation of Korea (2016R1A6A3A11932416), and the project with the title of Economics of Cryptocurrency funded by BlockchainOS.

References

1. Leidner, D.E., Jarvenpaa, S.L.: The information age confronts education: case studies on electronic classrooms. Inf. Syst. Res. **4**(1), 24–54 (1993)
2. Leidner, D., Jarvenpaa, S.: The use of information technology to enhance management school education: a theoretical view. MIS Q. **19**(3), 265–291 (1995). https://doi.org/10.2307/249596
3. Wastiau, P., Blamire, R., Kearney, C., Quittre, V., Van de Gaer, E., Monseur, C.: The use of ICT in education: a survey of schools in Europe. Eur. J. Educ. **48**(1), 11–27 (2013)
4. Sánchez, J.J.C., Alemán, E.C.: Teachers' opinion survey on the use of ICT tools to support attendance-based teaching. Comput. Educ. **56**(3), 911–915 (2011)
5. Tezci, E.: Factors that influence pre-service teachers' ICT usage in education. Eur. J. Teach. Educ. **34**(4), 483–499 (2011)
6. Sarkar, S.: The role of information and communication technology (ICT) in higher education for the 21st century. Science (80-) **1**(1), 30–41 (2012)
7. Nkenlifack, M., Nangue, R., Demsong, B., Fotso, V.K.: ICT for education. Education **2**(4), 30–39 (2011)
8. Dosi, G.: Technological paradigms and technological trajectories: a suggested interpretation of the determinants and directions of technical change. Res. Policy **11**(3), 147–162 (1982)

9. Baek, S., Kim, K., Altmann, J.: Role of platform providers in service networks: the case of salesforce. com app exchange. In: 2014 IEEE 16th Conference on Business Informatics (CBI), pp. 39–45. IEEE (2014)
10. Kim, K., Lee, W.-R., Altmann, J.: SNA-based innovation trend analysis in software service networks. Electron Mark. 25(1), 61–72 (2015)
11. McCroskey, J.C., Andersen, J.F.: The relationship between communication apprehension and academic achievement among college students. Hum. Commun. Res. 3(1), 73–81 (1976)
12. Frymier, A.B., Houser, M.L.: The teacher-student relationship as an interpersonal relationship. Commun. Educ. 49(3), 207–219 (2000)
13. Goldin, A.P., Pezzatti, L., Battro, A.M., Sigman, M.: From ancient Greece to modern education: universality and lack of generalization of the Socratic dialogue. Mind, Brain, Educ. 5(4), 180–185 (2011)
14. Gupta, S., Bostrom, R.: Research note—an investigation of the appropriation of technology-mediated training methods incorporating enactive and collaborative learning. Inf. Syst. Res. 24(2), 454–469 (2013)
15. Alavi, M., Wheeler, B., Valacich, J.: Using IT to reengineer business education: an exploratory investigation of collaborative telelearning. MIS Q. 19(3), 293–312 (1995). https://doi.org/10.2307/249597
16. Alavi, M.: Computer-mediated collaborative learning: an empirical evaluation. MIS Q. 18 (2), 159–174 (1994). https://doi.org/10.2307/249763
17. Ozdemir, Z.D., Abrevaya, J.: Adoption of technology-mediated distance education: a longitudinal analysis. Inf. Manag. 44(5), 467–479 (2007)
18. Söllner, M., Bitzer, P., Janson, A., Leimeister, J.M.: Process is king: evaluating the performance of technology-mediated learning in vocational software training. J. Inf. Technol. 33(3), 233–253 (2018)
19. Santhanam, R., Sasidharan, S., Webster, J.: Using self-regulatory learning to enhance e-learning-based information technology training. Inf. Syst. Res. 19(1), 26–47 (2008)
20. Jonassen, D.H., (eds.): Handbook of Research for Educational Communications and Technology. A Project of the Association for Educational Communications and Technology. Simon & Schuster Macmillan, New York (1996)
21. Xu, D., Huang, W.W., Wang, H., Heales, J.: Enhancing e-learning effectiveness using an intelligent agent-supported personalized virtual learning environment: an empirical investigation. Inf. Manag. 51(4), 430–440 (2014)
22. Webster, J., Hackley, P.: Teaching effectiveness in technology-mediated distance learning. Acad. Manag. J. 40(6), 1282–1309 (1997)
23. Wan, Z., Wang, Y., Haggerty, N.: Why people benefit from e-learning differently: the effects of psychological processes on e-learning outcomes. Inf. Manag. 45(8), 513–521 (2008)
24. Chang, V.: Review and discussion: e-learning for academia and industry. Int. J. Inf. Manag. 36(3), 476–485 (2016)
25. Aparicio, M., Bacao, F., Oliveira, T.: An e-learning theoretical framework. Educ. Technol. Soc. 19(1), 292–307 (2016)
26. Keleş, M.K., Özel, S.A.: A review of distance learning and learning management systems. In: Virtual Learning. InTech (2016)
27. Tomei, L.A.: Encyclopedia of Information Technology Curriculum Integration. IGI Global, Hershey (2008)
28. Piccoli, G., Ahmad, R., Ives, B.: Web-based virtual learning environments: a research framework and a preliminary assessment of effectiveness in basic IT skills training. MIS Q. 25(4), 401–426 (2001). https://doi.org/10.2307/3250989
29. Anohina, A.: Analysis of the terminology used in the field of virtual learning. J. Educ. Technol. Soc. 8(3), 91–102 (2005)

30. Zhang, D., Zhao, J.L., Zhou, L., Nunamaker Jr., J.F.: Can e-learning replace classroom learning? Commun. ACM **47**(5), 75–79 (2004)
31. Zhang, D., Zhou, L., Briggs, R.O., Nunamaker Jr., J.F.: Instructional video in e-learning: assessing the impact of interactive video on learning effectiveness. Inf. Manag. **43**(1), 15–27 (2006)
32. Straub, E.T.: Understanding technology adoption: theory and future directions for informal learning. Rev. Educ. Res. **79**(2), 625–649 (2009)
33. Mumtaz, S.: Factors affecting teachers' use of information and communications technology: a review of the literature. J. Inf. Technol. Teach. Educ. **9**(3), 319–342 (2000)
34. Tabata, L.N., Johnsrud, L.K.: The impact of faculty attitudes toward technology, distance education, and innovation. Res. High. Educ. **49**(7), 625 (2008)
35. Google Inc. Google Maps [Internet]. https://www.google.com/maps/place/Waco,+TX, +USA/@31.5573106,-97.2608857,11.75z/data=!4m5!3m4!1s0x864f82f1230151d3:0xfbd7 4b03d6d1aa10!8m2!3d31.549333!4d-97.1466695. Accessed 20 July 2018
36. Amazon Web Services Inc. Amazon EC2 [Internet]. https://aws.amazon.com/ec2/. Accessed 20 July 2018
37. Zang, N., Rosson, M.B., Nasser, V.: Mashups: who? what? why? In: CHI 2008 Extended Abstracts on Human Factors in Computing Systems, p. 3171–6. ACM (2008)
38. Oestreicher-Singer, G., Sundararajan, A.: Recommendation networks and the long tail of electronic commerce. Mis Q. 65–83 (2012)
39. Haile, N., Altmann, J.: Value creation in software service platforms. Futur. Gener. Comput. Syst. **55**, 495–509 (2016)
40. ProgrammableWeb [Internet]. https://www.programmableweb.com/. Accessed 21 July 2018
41. Mendeley Ltd. Mendeley Opens up Science for Everyone [Internet]. https://dev.mendeley. com/. Accessed 21 July 2018
42. Berkeley Univercity of California. API Central [Internet]. https://api-central.berkeley.edu/. Accessed 21 July 2018
43. Rogers, E.M.: Diffusion of innovations. Simon and Schuster (2010)
44. Bass, F.M.: A new product growth for model consumer durables. Manag. Sci. **15**(5), 215–227 (1969)
45. Jovanovic, B., MacDonald, G.M.: The life cycle of a competitive industry. J. Polit. Econ. **102**(2), 322–347 (1994)
46. Moore, G.A.: Crossing the Chasm: Marketing and Selling Disruptive Products to Mainstream Customers. Harper Collins, New York (2014)
47. Gawer, A., Cusumano, M.A.: Platform leadership: How Intel, Microsoft, and Cisco Drive Industry Innovation, vol. 5. Harvard Business School Press, Boston (2002)
48. Wasserman, S., Faust, K.: Social Network Analysis: Methods and Applications, vol. 8, pp. 233–238. Cambridge University Press, New York (1994)
49. Festinger, L.: A Theory of Cognitive Dissonance, vol. 2. Stanford University Press, Stanford (1962)
50. Newman, M.E.J.: Finding community structure in networks using the eigenvectors of matrices. Phys. Rev. E **74**(3), 36104 (2006)
51. Albert, R., Jeong, H., Barabási, A.-L.: The diameter of the world wide web. arXiv Prepr cond-mat/9907038 (1999)
52. Albert, R., Jeong, H., Barabási, A.-L.: Error and attack tolerance of complex networks. Nature **406**(6794), 378 (2000)
53. Peres, R., Muller, E., Mahajan, V.: Innovation diffusion and new product growth models: a critical review and research directions. Int. J. Res. Mark. **27**(2), 91–106 (2010)
54. Kim, J., Ilon, L., Altmann, J.: Adapting smartphones as learning technology in a Korean university. J. Integr. Des. Process Sci. **17**, 5–16 (2013)

Model and Simulation Engines for Distributed Simulation of Discrete Event Systems

José Ángel Bañares[(✉)] and José Manuel Colom

Computer Science and Systems Engineering Department,
Aragón Institute of Engineering Research (I3A), University of Zaragoza,
Zaragoza, Spain
{banares,jm}@unizar.es

Abstract. The construction of efficient distributed simulation engines for discrete event systems (DES) remains a challenge. The vast majority of simulations that are developed today are based on federation of modular sequential simulations. This paper proposes the steps to fill the gap from specifications based on Petri Nets to an efficient simulation of the net throughout a distributed application devoted to this purpose and exploiting the versatility of cloud infrastructures. The outcomes of the proposed DES distributed simulation are: (1) an adapted execution model of PN that is based in the generation and management of events related to the enabling and occurrence of transitions; (2) simple simulation engines for these adapted PN, each hosting a subset of transitions; (3) an scheme for deployment of a set of connected simulation engines; and (4) a simple mechanism for dynamic load balancing by merging/ splitting the subsets of transitions hosted in simulation engines.

Keywords: Distributed simulation · Discrete event systems ·
Dynamic load balancing · Petri Nets

1   Introduction

In many fields, ranging from healthcare monitoring to industrial manufacturing applications, the systems are becoming very large and complex. Moreover, they must be designed as part of an interconnected world. Smart systems (Cities, Buildings, Factories, Logistics) are examples of such systems. They share a set of characteristics such as: involvement of physical and computational interactions, integration of human behaviour into the processes, consideration of sustainability and economical requirements, and achievement of unprecedented levels of scale and complexity. The construction of models for these systems, that retain the essential elements and parameters for its design, is an accepted strategy to cope with these systems. Nevertheless, the modelling of these systems often gives rise to models that cannot be used in practice in the design, analysis or implementation processes. These problems arise because of the high-level semantics of

© Springer Nature Switzerland AG 2019
M. Coppola et al. (Eds.): GECON 2018, LNCS 11113, pp. 77–91, 2019.
https://doi.org/10.1007/978-3-030-13342-9_7

the models obtained or the size of the model, which makes the model unmanage-able for the available tools inside the engineering process oriented to design, the evaluation of architectural solutions, or the assessment of system performance.

For this kind of complex and scalable systems, simulation becomes the only alternative available in practice for the different tasks in its life cycle. In this case, it is an essential tool for system operation, to dynamically enable the continuous design, configuration, monitoring and maintenance of operational capability, quality, and efficiency. The capacity to trigger simulations in a short period of time to anticipate the effect of control actions is an essential tool to transform the high-volume of continuously streaming data into knowledge for decision support.

This paper is focused on discrete event Systems (DES), where the evolution from one system's state to another is produced as a consequence of the appear-ance of a *discrete relevant fact* for the system that is called *event*. The system's actions happen or are executed while the system is in a state and have an asso-ciated temporary duration, an economic cost, etc. The completion of a system action causes the state change of the system in an atomic manner. The simula-tion of a DES consists in the execution of a model that represents the system. This model must represent the state of the system and the state transitions that are the discrete state changes that may occur at discrete points in simulation time, and when an event occurs [33].

The introduction of acceleration techniques in simulation applications has been a permanent objective that has been strongly related with the growth of the size of the systems to be simulated. Parallelism and distribution are techniques oriented to this goal. However, to obtain simulation execution times better than in a centralized simulation it is necessary to consider a careful selection of the execution model to be used, the partition and distribution of the model to be simulated, and the analysis of message traffic between the simulation engines, which in general are closely coupled tasks.

Parallel and Distributed Discrete Event Simulations tools offer the ability to perform detailed simulations of large-scale computer networks [11], traffic [32], and military applications [29], among other applications. Despite the relevance of large-scale DES simulations, this type of problem is far from be solved and poses important challenges [12,14]. The difficulty to move these applications to the cloud can be exemplified by the modelling and simulation of the cloud itself [5]. A review of thirty-three cloud simulators is presented in [4], but just one of the tools reviewed *cloud2Sim* considers distributed simulations.

The main challenges of Distributed DES simulations pointed in [12] are:

- The definition of **modeling languages** allowing the generation of *efficient parallel and distributed simulation code*.
- The statement of a **clear execution semantic of the model**, and the **execution policy** of its interpreter [18,26]. They must be oriented to a distributed implementation.
- The availability of *load balancing mechanisms* to cope with the unpre-dictability of the underlying execution environment [7,8].

– The incorporation of the **economic cost** and **energy consumption**, in addition to the traditional speedup metric in distributed simulations.

Petri Nets (PNs) have been pointed out as a good formalism for modelling realistic features and perspectives of reactive and distributed systems, such as control flow, data, resources, and for analyzing and verifying many properties. The automatic analysis of properties is supported by software tools, and when formal analysis become impracticable, the model must be simulated. In this paper we focus on these challenges using PNs. Section 2 presents succinctly the work related to the distributed simulation of DES. Section 3 provides a global overview of the methodology. It covers the automatic translation of PN specifications in efficient parallel and distributed code, the impact of distribution on the execution policies, the definition of simple interpreters to support a distributed simulation, and the support for efficient load balancing between distributed simulators. In this paper we specialize the methodology presented in [30] for conceptual modeling of DES in distributed simulation. The methodology is supported by a high level PN (HLPN) based specification supporting modularity and hierarchy for the modeling of complex systems that was presented in [20]. In our previous work [19], we show how to translate a HLPN into a flat model by means of an elaboration process. In this paper we focus into the process to automatize the translation of flat PN models to efficient parallel simulation code. Finally, Sect. 4 presents an actor architecture for an efficient distributed simulation exploiting the PN execution representation.

2 Related Work

There has been a significant amount of work in the field of parallel and distributed DES simulation. A historical review can be found in [10], and many of the current challenges of the discipline has been recently collected in [12].

The discipline began defining *logical processes (LP)* and the *synchronization problem* with what is known as the Chandy/Misra/Bryant algorithm. Synchronization protocols and variants of conservative and optimistic approaches continues to be a focus of research to address synchronization and performance issues associated with executing parallel discrete event simulations in cloud computing [17].

The other focus of research is concerned with an architectural point of view, the development of middleware, frameworks and standards. The High Level Architecture (HLA) is an standard developed by the United State's Department of Defense to perform distributed simulations for military purposes that became an Open IEEE Standard, and has been adopted as the facto standard for federating simulations [31]. Dynamic balancing for HLA-based simulations remains a challenge [8].

Distributed computing programs do not have the same requirements as those of parallel DES programs and thus infrastructure must be specifically designed to support this simulation environment. In [13], Fujimoto et al. propose a master/worker architecture called *Aurora*. Cloud computing is focusing the research with the expectation that the development of Simulations as a Service will hide the difficulty of developing efficient parallel simulation and will made distributed simulation broadly accesible to all users [27].

Related with the use of PNs for simulating DESs, the translation of a system model expressed by a PN to an actual hardware or software system with the same behavior as the model is a PN implementation. Given a PN model of a DES, the *simulation* of the system can be done by *playing the token game*, i.e. by moving tokens when transitions are enabled. If a deterministic or stochastic time interpretation is associated to transitions – Timed PNs (TPNs) or Stochastic PNs (SPNs) –, the interpretation of the TPN or SPN yields, actually, a Discrete Event Simulation system.

The implementation of a PN can be classified as *compiled* or *interpreted*. The *compiled* implementation generates code whose behavior corresponds to PN evolutions, while an interpreted PN codifies the structure and marking as data structures used by one or more interpreters to make the PN evolve. Compiled implementations has been the option for the development of discrete event control systems [21, 25]. *Interpreted* implementations has the advantage of separating the model specification from the simulator, which provides a number of benefits summarized by Z. Zeigler in [33]: (1) The model is not wired with the simulator, which enables the portability of the model to other simulator and interoperability at a high level of abstraction; (2) Algorithms for distributed simulation can be presented independently of the model; and (3) Model complexity is related with the number of resources required to correctly simulate a model. All these benefits are related with requirements for a distributed simulation. The principle of separation of model and simulator remains the base for scaling resources according to the size of the model, workload balancing by moving parts of the model between distributed simulator engines, and reusing good well defined PDES algorithms.

There has been substantial work in the 1990's on distributed simulation of TPNs [1, 6, 9, 23, 28], which show the TPN formalism can contribute to the efficient implementation of distributed discrete event simulations thanks to the PN structure. These works focused on good partitioning algorithms based on the PN structure and synchronization algorithms. In these works, the TPN is decomposed into a set of LPs assuming a FIFO communication. The interface of the LP is defined by a subset of places, and arcs connecting with these places are replaced by communication channels. LPs interact exchanging time-stamped messages that represent token transfers, and each LP executes a simulation engine that implements the same simulation strategy to interpret the PN partition and to preserve causality with events simulated by other LPs. However, these approaches do not consider the automated translation of the PN structure to efficient code for simulation engines.

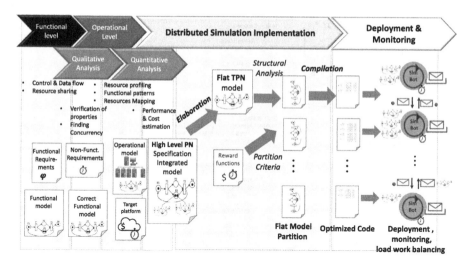

Fig. 1. PN-based process from specification to distributed model execution.

3 Event Driven Simulation Based on Petri Nets

3.1 The Overall Methodological Approach

In this section we present an approach to automatize the translation of TPN specifications into efficient parallel and distributed simulation code. The proposed steps are part of a methodological approach to manage the complexity of developing the logic of a complex system taking into account functional and not functional requirements, and gradually incorporating restrictions imposed by the underlying hardware infrastructure that was presented in [30]. Figure 1 depicts some steps of the methodology: *Functional models of systems* are built focusing on a set of concurrent communicating processes competing for shared resources. *Qualitative analysis* checks the model and help to find maximum concurrency. The *operational model* enriches the model with characteristics of the execution platform to develop a *quantitative analysis*. This analysis provides information useful for the partition of the model providing metrics required for the distributed execution. Figure 1 focuses on the last steps:

Elaboration. The objective of the first steps is the modelling of complex and large scale DES. It requires a formal description of different facets supported by a hierarchical and component decomposition. Modular and hierarchical PN specification, such as introduced in [20], provides more compact and manageable descriptions. However, the interpretation of a high level model can introduce important sources of inefficiency due to higher levels of abstraction such as efficient matching to evaluate enabled transitions [22], or introduce complex synchronization protocols in distributed simulations. Instead of the direct emulation of high level models, we propose transform the original model to be simulated

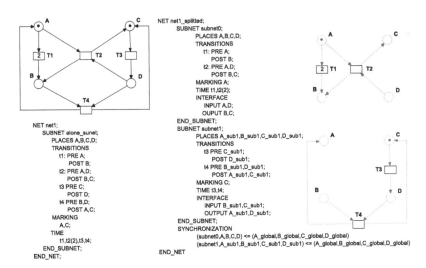

Fig. 2. Graphical, textual and splitted textual specifications of a TPN.

into a flat Place/Transition net model. This transformation process called *elaboration* was illustrated in [19] with the elaboration of a high level PN to a flat model of sequential state machines.

Compilation. The compilation stage transforms a flat Place/Transition net into executable code/data. A classical PN simulation engine follows a repetitive cycle that involves: (a) to scan enabled transitions; (b) to fire some of the detected enabled transitions (executing, maybe, some associated activity), and (c) to update the marking (the state) of the PN. Although the elaboration process simplifies the complexity of the enabling tests of transitions by removing the need of unification algorithms, the enabling test in a Place/Transition net remains to consume most of the interpreter loop.

Distribution. Partitioning requires to proceed, a priori, identifying the *good* subnets in which the original one is divided. The initial model partition can be supported by applying structural and behavioral analysis [15], in this sense strategies based into the identification of sequential state machines can be used (computing for example p-semiflows in an incremental way). Alternative partition approaches can be found in [1,21]. The hardware architecture and syncronization algorithms can be taken into account [2,7,19].

Load Balancing. Thanks to a simulation based on identical simulation engines working on data structures and variables representing PNs, it is possible to make a dynamic reconfiguration of the initial partition: (1) by fusion of the data structures of two simulation engines in only one; or (2) by splitting the data structure contained in a simulation engine into two separate data structures over two distinct simulation engines. This dynamic reconfiguration is not possible in simulation contexts where the system to be simulated is not a data structure (e.g. the system is a program that must be compiled).

3.2   Simulation of PNs Oriented to Distributed Implementation

We can compare the interpretation of PN simulation engines with the interpretation of rule based systems (RBS). Both describe how a model evolves in time. RBS take advantage of *temporal redundancy* based on the idea that most of data in memory does not change when a rule is fired in each interpretation cycle, and most of rules remains enabled or not enabled. Based on this idea, a compilation process builds a (RETE) network that connects state changes with rules affected by state changes, and store partial matching operations. An adaptation of the RETE network for the centralized interpretation of HLPNs was proposed in [22]. It is possible to go beyond improving the efficiency of the PN interpreter by (1) removing complex matching operations in the elaboration process; (2) replacing them by simple linear functions; and (3) incorporating postconditions to the compilation process. In RBS postconditions are left out of the compilation process due to postconditions produces data modifications that can not be related with state changes. However PNs explicitly specify preconditions and state changes giving the possibility of compiling them in a network. This approach is followed in [3] defining the so called *Linear Enabling Function* (LEF) of a transition in Place/Transition specifications that allow to characterize when a transition is enabled (can occur).

We propose in this section the translation of the structure and marking of a Place/Transition net to a set of LEFs as optimized code for distributed simulation engines. The LEF of a transition t, $f_t : \mathbf{R}(\mathcal{N}, \mathbf{m_0}) \longrightarrow \mathbb{Z}$, maps each marking $\mathbf{m}$ belonging to the set of reachable markings, $\mathbf{R}(\mathcal{N}, \mathbf{m_0})$, to an integer, in such a way that t can occur for $\mathbf{m}$, iff $f_t(\mathbf{m}) \leq 0$. For example, for transition $T2$ in the net of Fig. 2, its LEF is: $f_{T2}(\mathbf{m}) = 2 - (\mathbf{m}[A] + \mathbf{m}[D]), \forall \mathbf{m} \in \mathbf{R}(\mathcal{N}, \mathbf{m_0})$, where $\mathbf{m_0}$ is the initial marking depicted in Fig. 2 (places A, C marked with a token and the rest of places unmarked). Observe that at $\mathbf{m_0}$, the value of the LEF is $f_{T2}(\mathbf{m_0}) = 1 > 0$, i.e. the transition $T2$ is not enabled at $\mathbf{m_0}$. Nevertheless, at the reachable marking, $\mathbf{m}$, that contains one token in place A and one token in D, $f_{T2}(\mathbf{m}) = 0$ indicating that the transition $T2$ is enabled and can occur.

The use of LEFs, as presented before for the characterization of the enabling of a transition, requires a explicit representation of the marking of the net and the LEF itself as a function. For a distributed simulation this gives rise to two problems that make this execution model of a PN not well-adapted for this purpose: (1) The explicit representation of the marking in a distributed environment is a set of shared variables between a set of distributed simulation engines that requires mechanisms for the maintenance of coherence and consistency of the marking variables (this is is in fact a bottleneck for distributed simulation); (2) The funcional representation of the LEF requires its continuous evaluation for the marking in order to determine the enabling of a transition.

To address these two problems, the LEF mechanism for a transition is implemented according to the following principles: (1) Only the current value of the LEF (initially this value corresponds to the value of the LEF at $\mathbf{m_0}$ and computed in compilation time) is stored; (2) Each time a transition occurs in the net, a constant is sent to each transition which enabling has been affected by the occurrence of the transition. This constant is used for the updating of the LEF of the affected transition.

With this strategy, the explicit representation of the marking and the re-evaluation of the LEF are not needed. The changes of a LEF are based in the constants sent by the transitions that modify its enabling conditions. This is the reason why this execution model for PN's simulation becomes a Discrete Event Simulation because the events are the constants sent by the occurrence of a transition to all transitions whose enabling conditions have been changed by its occurrence, and the simulation state becomes the current values of LEFs.

That is, if $\mathbf{m} \xrightarrow{t'} \mathbf{m}'$, $f_t(\mathbf{m}')$ can be computed from the value of $f_t(\mathbf{m})$ and a static parameter known at compilation time that represents the change of f_t after the occurrence of t'. This parameter corresponds to changes in the contents of tokens of the input places of t as a consequence of the occurrence of t'. Thus, the updating equation for any LEF when t' occurs takes the form $f_t(\mathbf{m}') = f_t(\mathbf{m}) + UF(t' \longrightarrow t)$, where $UF(t' \longrightarrow t)$ is known as the *Updating Factor* of t' over t obtained from the structure of the net and its initial marking.

According to the previous comments, the compilation of a Place/transition net produces a representation of the net where there is an entry for each transition *(information of the LEF mechanism)*, t, grouping: (1) The variable maintaining the current value of the LEF and initialized to $f_t(\mathbf{m_0})$; and (2) The list of updating factors (simulation's events), $UF(t \longrightarrow t')$, that will be sent to each $t' \in (\bullet t)^\bullet \cup (t^\bullet)^\bullet$ whose enabling conditions have been affected by the occurrence of t. Observe that the partition of a model for a distributed simulation only requires to define the set of transitions to be grouped in each one of distributed simulation engines and load the previous data associated to the transitions in the corresponding engine. The information associated to each transition corresponding to the described LEF mechanism is independent to the information of any other transition. So, in order to perform the dynamic load balancing of the simulation's workload, it is enough to move from one engine to another the information of the LEF mechanism of the transitions to be moved. See Fig. 3.

The kernel of each distributed simulation engine to implement the Discrete Event Simulation of the Petri Net, according to the execution model based on the LEF mechanism described before, essentially: (1) Updates the LEFs of transitions with the updating factor interchanged; (2) Scans the list of the variables containing the current values of the LEFs in order to detect the enabled ones (values less than or equal to 0); and (3) Proceeds to make all the operations corresponding to the occurrence of the enabled transitions, executing the associated actions, and sending the list of Updating Factors stored together the value of the LEF. Figure 4 presents an algorithm that implements this kernel of the basic simulation engine [3], using the following information associated to each transition, t', belonging to the part of the PN model to be simulated (See Fig. 3): (1) **Identifier** of t'. A *global name* recognised in all sites of the simulation process; (2) $\tau(t')$. **Deterministic firing time** associated to transition t'. It stands for the duration time of the action associated to the occurrence of t'; (3) **Counter.** Variable containing the current value of the LEF $f_{t'}(\mathbf{m})$, initialized

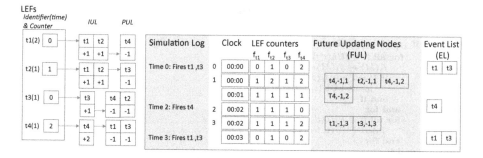

Fig. 3. Compilation result for the PN in Fig. 2 using the LEF mechanism.

with $f_{t'}(\mathbf{m_0})$, and updated whenever the transition –or a transition affecting it– occurs, according to the received Updating Factor; (4) **Immediate Updating List** $(IUL(t'))$ Set of transitions $(\bullet t')\bullet$ whose LEFs must be updated after the occurrence of t' containing the corresponding Updating Factor to be sent (Note that $(\bullet t')\bullet$ includes t'); (5) **Projected Updating List** $(PUL(t'))$ Set of transitions $(t'\bullet)\bullet$ whose LEFs must be updated after the occurrence of t' containing the corresponding Updating Factor to be sent.

The algorithm in Fig. 4 receives the *LEFs*, a list of transition nodes representing the PN to simulate for the simulation engine, and the limit of the virtual time to be simulated. *EL* contains enabled transitions. *FUL* contains *Future Updating Nodes* (FUNs), and the function *insert-FUL()* maintains them ordered by time. *FUL* plays the role of the Future Event List in an event-driven simulation algorithm for DES. A FUN holds: a pointer (pt) to the transition to be updated, the updating factor $UF(t' \to t)$ delivered by each fired transition $(t \in (t'\bullet)\bullet)$, and the time $(time)$ at which the updating must take effect. *head-FUL* is a pointer to *FUL*, *pop(FUL)* pops and returns the head of *FUL*, and we access the fields of *FUNs* using the dot notation. The variable *clock* holds the current simulation time. Figure 3 shows transition nodes in the previously presented PN.

Observe that the interpreter immediately applies *IUF* updating factors, which represents removing tokens from previous places, once a transition occurs, but insert events in *PUL*, which represent that tokens will be appear in posterior places at future clock time. It is important to note also that the interpreter takes all enabled transitions in the *EL* in order, solving in this way conflicts. This execution policy avoids the state with tokens simultaneously in place A and D, which is a possible state. A random number of enabled transitions can be taken in each interpretation cycle, or updating factors in *IUL* and *PUL* can be atomically applied, representing and atomic occurrence of transitions. This alternative implementations suppose alternative execution policies that can result in different executions. To avoid it, beside the execution policy specification, it is required to identify transitions in conflict and the policy to solve them.

Finally, it is important to point out that the execution model based on the LEF mechanism makes unnecessary the representation and updating of the

```
 1: procedure SIMULATE(Lefs, EL, simulationTime)
 2:     VT ← 0; FUL ← {};
 3:     for all (t' ∈ EL) do                                        ▷ Fires enabled transitions
 4:         if (f_{t'}(M) ≤ 0) then                                 ▷ Checks transition is enabled yet
 5:             for all (t ∈ IUL(t')) do
 6:                 f_t(M) ← f_t(M) + UF(t' → t);
 7:                 if (t = t' and f_t(M) ≤ 0) then                 ▷ Avoids race conditions
 8:                     insert-FUL (t, 0, τ(t) + clock);
 9:                 end if
10:             end for
11:             for all (t ∈ PUL(t')) do
12:                 insert-FUL (t, UF(t' → t), τ(t') + clock);
13:             end for
14:         end if
15:     end for
16:     if (head-FUL.time > clock) then VT ← head-FUL.time          ▷ Update Virtual Time
17:     end if
18:     while (head-FUL.time = VT) do                               ▷ Update Event List
19:         t ← head-FUL.pt; f_t(M) := f_t(M) + head-FUL.UF;
20:         if (f_t(M) ≤ 0) then insert(EL, t);
21:         end if
22:         head-FUL ← pop(FUL);
23:     end while
24: end procedure
```

Fig. 4. Centralized simulator engine.

marking of the PN model. Nevertheless, the construction of the marking of the PN after the occurrence of a sequence of transitions can be easily done collecting a log containing the occurrence of transitions each one labelled with the simulation time. From this log of labelled transitions, the occurrence sequence can be reconstructed and then using the net state equation (an algebraic computation), for example, compute the reached marking from the initial one.

4   A Framework for Distributed Simulations of DESs

Distributed simulation of TPNs will be based on many identical simulation engines distributed over the execution platform, and each one devoted to the simulation of a subnet of the original one. Each subnet is represented in the corresponding simulation engine as a data structure. In [24], K.S. Perumalla points out the need of micro-kernels specialized in simulation for building distributed simulations. The idea is to have a micro-kernel that collects the core invariant portion of distributed DES simulation techniques, and avoids to develop entirely the systems from scratch. The core must permit traditional implementations (conservative or optimistic), and the incorporation of newer techniques. We will call *SimBots* to our micro-kernels implementing LP.

The simulation engine proposed by Chiola and Ferscha in [6] manage subnet regions as a subset of places, transitions, and arcs of the original net. This implies that the interface is defined by a subset of places. Moreover, in [6] conservative and optimistic approaches assume a communication channel for each arc connecting the corresponding TPN regions. Therefore, a dynamic workload balancing would require a continuous interface redefinition configured by different channels, which is a very big problem. The availability of a scalable architecture for large-scale simulations requires and event driven execution model. The actor model based on asynchronous message passing has been selected for the design of large scale distributed simulations as unit of concurrency [16]. It is an event driven model that scales to a large number of actors and removes the complexity of locking mechanisms. A single immutable interface that consists of a mailbox that buffers incoming messages, and a pattern based selection of messages to process them provides the flexibility for configuring different partitions. A distributed simulation based on the LEF mechanism requires a set of simple simulation engines called *SimBot* that each one can be considered as an actor. A *SimBot* will have a mailbox interface, an execution kernel based on the LEF mechanisms, and where is possible to implement different services: synchronization strategies, PN interpretations, load balancing redistributing transition nodes, and self-configuring partition strategies according to the state of computational and network resources.

A *SimBots'* system can be seen in Fig. 5. Transition nodes configure a network that processes events triggering updating factors of adjacent transitions when the guard representing the counter equals or less than zero. To distribute the network, it is only required to route messages to the SimBot that contains the corresponding transition nodes. It can be easily done with a *transition service name*, or using routing services supported by actor models such as Akka. Figure 5 shows in the wall clock time axis how evolves the system. Initially the *Simulator System* receives a textual specification of a flat PN, and it is sent to the PNcompiler to obtain the *LEF data structure*. Figure 2 shows how an initial PN specification can be split in subnets. An initial criteria can be as simple as avoid the distribution of transitions in conflict, which can be obtained by structural analysis and to balance the number of transitions in subnets. The *monitoring & load balancing* actor deploys compiled code and monitors the simulation state.

Once the code is deployed, *SimBots* can interchange asynchronous messages with time-stamped updating factors. Each *Simbot* execute the same strategy, incorporating events of adjacent SimBots to the interpretation loop. The *Monitoring & load balancing* actor can recover logs (list of time-stamped triggered transitions). By joining and ordering events it can be obtained a global consistent state and it is possible to monitor the simulation. The bottom part of the figure shows how the system can perform workload balancing as the result of a self-configuration of adjacent actors. The compilation process also incorporates structural information to know adjacent actors, that is, *SimBots* that send or receive updating factors. In the load balancing process, the *SimBot* must be sychronized with adjacent *SimBots* until the LEF data structure corresponding to transitions can be moved. The set of transitions hold by a *SimBot* can be split to distribute it between adjacent *SimBots*, or can be joined in a *SimBot* resulting in a inactive one if there is not transitions to deal.

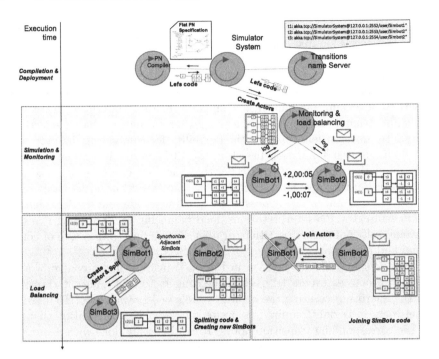

Fig. 5. An actor-based architecture for distributed PN simulations.

5   Conclusions and Future Work

A process to fill the gap between high level specification of complex DESs and the generation of code for scalable and dynamic distributed simulations has been presented. The process is based on the well known formalism of PNs, and it is presented an efficient representation for its interpretation. The codification lacks of state representation and makes easy load balancing between interpreters. From the performance point of view, the simulation technique of PNs presented here is the method *Enabled Transitions* (ET) introduced in [25], where places are all 1-bounded. In this paper, it is shown that in a centralized environment ET is better than other simulation techniques when the number of processes grows above a threshold. If we are concerned with distributed simulations, the cost of interchanged messages must be also considered. In general, the transmission of constants after transition firing must be considered in all methods, but in our method we don't need messages related to the maintenance of a global and consistent state of the full model required by other methods.

An actor architecture for distributed simulation of PNs has been also presented. Currently a prototype has been developed in Akka, with a compiler for simple binary PNs, and a basic *SimBot* actor able to interpret and transfer LEF data structures to adjacent *SimBots*. The execution model presented here and the mechanisms for its distributed implementation are the core for the development of any simulation strategy, e.g. conservative, optimistic, etc. It will be the basis for the exploration of new synchronization algorithms, self-configuring policies, and the definition of complex partition criteria considering economical aspect that will allow the development of Simulation as a Service.

Acknowledgments. Work financed by the Aragonese Government and the European Regional Development Fund "Construyendo Europa desde Aragón".

References

1. Ammar, H.H., Deng, S.: Parallel simulation of stochastic petri nets using spatial decomposition. In: IEEE International Symposium on Circuits and Systems, vol. 2, pp. 826–829, June 1991
2. Boukerche, A.: An adaptive partitioning algorithm for distributed discrete event simulation systems. J. Parallel Distrib. Comput. **62**(9), 1454–1475 (2002)
3. Briz, J.L., Colom, J.M.: Implementation of weighted place/transition nets based on linear enabling functions. In: Valette, R. (ed.) ICATPN 1994. LNCS, vol. 815, pp. 99–118. Springer, Heidelberg (1994). https://doi.org/10.1007/3-540-58152-9_7
4. Byrne, J., et al.: A review of cloud computing simulation platforms and related environments. In: CLOSER (2017)
5. Calheiros, R.N., Ranjan, R., Beloglazov, A., De Rose, C.A.F., Buyya, R.: Cloudsim: a toolkit for modeling and simulation of cloud computing environments and evaluation of resource provisioning algorithms. Softw. Pract. Exper. **41**(1), 23–50 (2011)
6. Chiola, G., Ferscha, A.: Distributed simulation of petri nets. IEEE Concurrency **3**, 33–50 (1993)
7. D'Angelo, G., Marzolla, M.: New trends in parallel and distributed simulation: from many-cores to cloud computing. Simul. Model. Prac. Theory **49**, 320–335 (2014)
8. De Grande, R.E., Boukerche, A.: Dynamic balancing of communication and computation load for HLA-based simulations on large-scale distributed systems. J. Parallel Distrib. Comput. **71**(1), 40–52 (2011)
9. Djemame, K., Gilles, D.C., Mackenzie, L.M., Bettaz, M.: Performance comparison of high-level algebraic nets distributed simulation protocols. J. Syst. Archit. **44**(6–7), 457–472 (1998)
10. Fujimoto, R.M., et al.: Parallel discrete event simulation: the making of a field. In: 2017 Winter Simulation Conference (WSC), pp. 262–291, December 2017
11. Fujimoto, R.M., Perumalla, K., Park, A., Wu, H., Ammar, M.H., Riley, G.F.: Large-scale network simulation: how big? how fast? In: Proceedings MASCOTS 2003. 11th IEEE/ACM International Symposium Modeling, Analysis and Simulation of Computer Telecommunications Systems, pp. 116–123, October 2003
12. Fujimoto, R., Bock, C., Chen, W., Page, E., Panchal, J.H. (eds.): Research Challenges in Modeling and Simulation for Engineering Complex Systems. SFMA. Springer, Cham (2017). https://doi.org/10.1007/978-3-319-58544-4
13. Fujimoto, R., Park, A., Huang, J.C.: Towards flexible, reliable, high throughput parallel discrete event simulations. In: Ince, A.N., Bragg, A. (eds.) Recent Advances in Modeling and Simulation Tools for Communication Networks and Services, pp. 257–278. Springer, Boston (2007). https://doi.org/10.1007/978-0-387-73908-3_13

14. Fujimoto, R.M., Malik, A.W., Park, A.: Parallel and distributed simulation in the cloud. SCS M&S Mag. **3**, 1–10 (2010)
15. García, F., Villarroel, J.: Decentralized implementation of real-time systems using time petri nets. application to mobile robot control. In: IFAC Proceedings, vol. 31(4), pp. 11–16 (1998)
16. Haller, P.: On the integration of the actor model in mainstream technologies: the scala perspective. In: Proceedings of the 2nd Edition on Programming Systems, Languages and Applications Based on Actors, Agents, and Decentralized Control Abstractions AGERE! 2012, pp. 1–6. ACM, New York (2012)
17. Malik, A., Park, A., Fujimoto, R.: Optimistic synchronization of parallel simulations in cloud computing environments. In: IEEE International Conference on Cloud Computing CLOUD 2009, pp. 49–56. IEEE (2009)
18. Marsan, M.A., Balbo, G., Bobbio, A., Chiola, G., Conte, G., Cumani, A.: The effect of execution policies on the semantics and analysis of stochastic petri nets. IEEE Trans. Soft. Eng. **15**(7), 832–846 (1989)
19. Medel, V., Arronategui, U., Bañares, J.Á., Colom, J.-M.: Distributed simulation of complex and scalable systems: from models to the cloud. In: Bañares, J.Á., Tserpes, K., Altmann, J. (eds.) GECON 2016. LNCS, vol. 10382, pp. 304–318. Springer, Cham (2017). https://doi.org/10.1007/978-3-319-61920-0_22
20. Merino, A., Tolosana-Calasanz, R., Bañares, J.Á., Colom, J.-M.: A specification language for performance and economical analysis of short term data intensive energy management services. In: Altmann, J., Silaghi, G.C., Rana, O.F. (eds.) GECON 2015. LNCS, vol. 9512, pp. 147–163. Springer, Cham (2016). https://doi.org/10.1007/978-3-319-43177-2_10
21. Moreno, R.P., Tardioli, D., Salcedo, J.L.V.: Distributed implementation of discrete event control systems based on petri nets. In: Proceedings of IEEE International Symposium Industrial Electronics, pp. 1738–1745, June 2008
22. Muro-Medrano, P.R., Bañares, J.A., Villarroel, J.L.: Knowledge representation-oriented nets for discrete event system applications. IEEE Trans. Syst. Man Cybern. Part A **28**(2), 183–198 (1998)
23. Nicol, D.M., Mao, W.: Automated parallelization of timed petri-net simulations. J. Parallel Distrib. Comput. **29**(1), 60–74 (1995)
24. Perumalla, K.S.: μsik - a micro-kernel for parallel/distributed simulation systems. In: Workshop on Principles of Advanced and Distributed Simulation (PADS 2005), pp. 59–68, June 2005
25. Piedrafita, R., Villarroel, J.L.: Performance evaluation of petri nets centralized implementation. the execution time controller. Discrete Event Dyn. Syst. **21**(2), 139–169 (2011)
26. Schriber, T.J., Brunner, D.T., Smith, J.S.: How discrete-event simulation software works and why it matters. In: Proceedings of the Winter Simulation Conference WSC 2012, pp. 3:1–3:15 (2012)
27. Shekhar, S., Abdel-Aziz, H., Walker, M., et al.: A simulation as a service cloud middleware. Ann. Telecommun. **71**(3), 93–108 (2016)
28. Thomas, G.S., Zahorjan, J.: Parallel simulation of performance petri nets: extending the domain of parallel simulation. In: 1991 Winter Simulation Conference Proceedings, pp. 564–573, December 1991
29. Tolk, A.: Engineering Principles of Combat Modeling and Distributed Simulation, 1st edn. Wiley, Hoboken (2012)
30. Tolosana-Calasanz, R., Bañares, J.Á., Colom, J.M.: Model-driven development of data intensive applications over cloud resources. Futur. Gener. Comput. Syst. **87**, 888–909 (2018)

31. Topçu, O., Durak, U., Oğuztüzün, H., Yilmaz, L.: Distributed Simulation: A Model-Driven Engineering Approach. Simulation Foundations, Methods and Applications. Springer, Cham (2016). https://doi.org/10.1007/978-3-319-03050-0
32. Zehe, D., Knoll, A., Cai, W., Aydt, H.: SEMSim cloud service: large-scale urban systems simulation in the cloud. Simul. Model. Pract. Theory **58**, 157–171 (2015)
33. Zeigler, B.P., Praehofer, H., Kim, T.G.: Theory of Modeling and Simulation: Integrating Discrete Event and Continuous Complex Dynamic Systems. Academic press, San Diego (2000)

An HVAC Regulation Architecture for Smart Building Based on Weather Forecast

Hanna Kavalionak[1]([✉])[iD] and Emanuele Carlini[2][iD]

[1] Dipartimento di Matematica e Informatica, University of Florence, Florence, Italy
hanna.kavalionak@unifi.it
[2] Istituto di Scienza e Tecnologie dell'Informazione (ISTI), National Research Council (CNR), Pisa, Italy
emanuele.carlini@isti.cnr.it

Abstract. Indoor climate control is one of the most important operations affecting the level of comfort, power consumption and costs in large buildings. The imminent proliferation of smart buildings equipped with a plethora of sensors and devices is a strong motivation to employ efficient and possibly automatic mechanisms to control indoor building climate. This paper proposes an high- level conceptual architecture for climate control in smart buildings, which is built on top of various state of the art approaches and solutions from different research fields. The core components of the architecture is heat transfer model to predict indoor temperature, which takes into account weather forecast and information coming from indoor sensors. The model is designed such that to adapt to different configurations and structural properties of buildings. The ultimate vision is the creation of a comprehensive system for indoor building temperature regulation to reduce energy consumption and operational costs of buildings without affecting (or even improving) the comfort conditions of its occupants.

1 Introduction

The long lasting vision of Smart Cities is the promise of an improved way of living urban environment, safer, comfortable and aware of the planet resources. Today many urban environments are migrating toward the implementation of smart features, including smart buildings. This migration is powered by the recent progresses in fields such as Internet of Things (IoT) and Cyber-Physical Systems (CPS), which provide the proper hardware and algorithms to implement self-regulation functions aimed at improving the quality of people life. This trend is motivated by the fact that people spend most of their time in buildings, for working, entertainment or residential purposes [15].

Many aspects of smart building are subject of extensive research, such as their surveillance [12, 13] or their energy efficiency [28]. One of the most critical aspect in a building is the control of the indoor climate (e.g. temperature and humidity),

© Springer Nature Switzerland AG 2019
M. Coppola et al. (Eds.): GECON 2018, LNCS 11113, pp. 92–103, 2019.
https://doi.org/10.1007/978-3-030-13342-9_8

usually performed by the so called Heating, Ventilation and Air Conditioning (HVAC) systems. This operation impacts on the people comfort as well as the power consumption for the building and, as a consequence, on management cost. As a matter of fact, HVAC functioning is the main energy-consuming factor in a building [20].

Simple degrees of automation are commonly found in widely available commercial systems, in which the scheduling of HVACs are operated by sensors when temperature drops below or above a given threshold. The functioning of the HVAC systems in large spaces is planned according to the hours of utilization, for example during the day for a typical offices environment. For example, the HVAC system starts few hours before the expected start of the working day and stops few hours before the end of the working day. This behaviour is based on the assumption that "few hours" are enough for heating/cooling the working space to a comfortable temperatures. This approach is often based on factual evidence and is far from an optimized way to adapt the temperatures inside the premises of the building. For example in the days following the week-ends, or in particular cold/warm days, the optimal number of hours can be very different.

The above reasons have motivated large research efforts in terms of automation and optimization of HVAC systems, spanning from physical models [2] to machine learning approaches [21]. More advanced approaches consider forecast about the outside temperature to schedule the system [6]. Considering the weather forecast allows not only to guarantee indoor climate to be comfortable, but also to reduce the energy consumption. For example in case the predicted outside temperature in the night hours is low, it is possible to schedule to use the outside air (e.g. by opening the windows) for cooling the inside. Further, in order to create an effective HVAC scheduling, one has to know not only the temperature inside the building in a given moment but also be able to predict its changes over time.

Following the above considerations, this paper proposes a design based on the unification of various state of the art approaches and solutions in different fields toward a comprehensive, conceptual architecture for HVACs regulation in the context of smart buildings. Please note that the purpose of this paper is not to provide a full architectural description of a real system, but rather representing a first step toward the identification of major components of a complex HVAC scheduling system. The architecture is built around a heat transfer model of an indoor space, derived by the composition of various state of the art approaches in thermodynamics. We show how to adjust the coefficients of such model to make it agnostic with respect to the structural composition of the building, when considering the indoor temperature. The model takes into account multiple heat sources affecting the building (i.e. conduction, radiation and other artificial heating/cooling sources) and the weather forecast. The purpose of the heat transfer model is to generate a prediction of the indoor climate. Such prediction can be used to feed decision making algorithms that organize the scheduling of HVAC according to the policy submitted by building administrators, including energy saving and cost reduction.

The paper is organized as the following. Section 2 briefly describes the relevant works in the context of HVAC regulation. The unified architecture is described in Sect. 3. Section 4 describes in details the heat transfer model. In Sect. 5 we perform a brief discussion on usual strategies for HVAC scheduling. Section 6 concludes the paper.

2   Related Work

Research work on HVAC regulation spans a multitude of topics, from physical modelling to optimization algorithm. In this section, we limit our brief review to approaches that are relevant from an architectural point of view in the context of smart buildings.

There are several studies about the prediction of energy consumption for building, which are based on human habits and behaviour [23, 28]. These works are the basis for research efforts that target indoor user comfort but considering the energy consumption and costs [3, 22]. For example, the work of Marche et al. [17] proposes an approach, contextualized in the Internet of Things (IoT), that studies the trade-off between the energy consumption and the comfort of the consumers for HVAC systems inside a building. The work [11] presents a smart heating and air conditioning scheduling method for a Home Energy Management System (HEMS). The proposed approach considers not only the occupants comfort, but also the characteristics of thermal appliances in the considered environment.

Several research works investigate the correlation between meteorological data and energy consumption. Di Corso et al. [4] propose a study on correlations between energy consumption and weather conditions at different granularity levels. This correlation inspired a number of research effort in integrating weather forecast into smart HVAC regulation systems. Du et al. [6] proposes a method that allows to adjust the actual energy consumption according to the error of the weather forecast.. Reynolds et al. [21] apply zone-level artificial neural networks in order to optimize the energy consumption in the building, using weather, occupancy and indoor temperature as input parameters.

Many HVAC scheduling systems are framed as Model Predictive Control (MPC) methods, in the tentative to represent the behavior of complex dynamical HVACs [14]. For example, Oldewurtel et al. [19] apply MPC together with weather predictions in order to increase energy efficiency of a building while respecting the occupant comfort. However, MPC-like algorithms have often high hardware and software requirements paired with complicated mechanisms for handling errors. As alternative, Drgoňa et al. [5], instead of directly applying MPC algorithms, introduce a framework for the generation of simplified control strategies that mimic an MPC method.

This paper builds upon several of the works mentioned above, as it proposes the unification of various state of the art approaches and solutions in different fields into a single, high-level architecture for HVACs regulation in the context of smart buildings. In particular, in order to model the HVAC system we propose to apply an hybrid approach, where the physics-based models are assisted by data-driven ones. In other words, the basic processes in HVAC are described by the physics-based models while the model coefficients are determined using the parameter estimation algorithms and the data collected from the indoor and outdoor sensors.

3 Architectural Overview

The main goal of an effective indoor climate regulation system for smart buildings is to maintain the desired indoor conditions in given time period while minimizing other factors, such as the power consumption and, consequently, the operative costs. To perform this task, an effective systems should consider both the current indoor conditions and the fluctuation of the outdoor weather, in such a way to optimize the working schedule of the various HVAC subsystem in the building.

The design of our proposed architecture (depicted in Fig. 1) revolves around this principle. The main idea is to exploit the indoor data and the outdoor weather data to adapt a *heat transfer model* for the building [7,9,11].

The architecture is composed by four main conceptual modules, namely: building and weather sensing, modelling & prediction, strategy planner, and HVAC control. In a real installation of the architecture, to each of the above functions can be associated to one or more hardware or software modules; this association depends on the specific building and set of constraints considered; This association is not further explored in this paper.

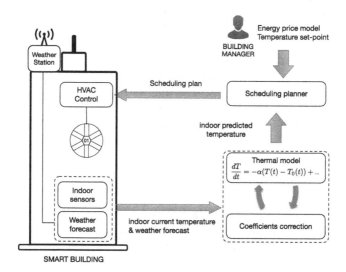

Fig. 1. Conceptual architecture of an unified HVAC regulation systems for smart buildings

The *building and weather sensing* module embeds the mechanisms to collect data about the outdoor and indoor climate of the building. In particular, we assume one or more weather stations are locally installed in the building to constantly monitor the outdoor climate, such as temperature, humidity and sun exposition. Similarly, indoor sensors measure the climate inside the building, including temperature and humidity. An additional sub-module is devoted to compute weather forecast and store past weather conditions.

The sensing module feeds the *model & prediction* module, whose main objective of is to predict the resulting indoor climate of the building, considering the data from the sensor and weather forecasts. The module exploits a heat transfer model of the building, which provides an estimation for the indoor climate in function of the time. A common challenge when implementing a heat transfer model for an actual building is the fine-tuning process, which allows to adapt to the structural properties of the building. In fact, the characteristics of the materials and the building orientation are not always easy to investigate but can significantly influence on the model behaviour. In order to overcome the necessity to know the construction particularities of the building we propose to apply a certain degree of self-learning in the model. Specifically, from the real data received from weather station and indoor sensors, we propose a mechanism to correct the heat transfer model coefficients (the approach is described in Sect. 4.3 in more details).

The *strategy planner* module exploits the above prediction to perform and actual scheduling planning for the HVAC system in the building. This module also considers the indications coming from the building administrators. Those can be about the desired comfort conditions, expressed as indoor temperature (i.e set-point) and humidity in function of time (e.g. working hours) and the energy provider price plan according to the time and the amount consumed. The output of this module is the definition of an efficient HVAC scheduling, which is described in more details in Sect. 5.

The *HVAC control* module represents the installed HVAC units in the building. We assume the existence of an unified control unit that is able to coordinate all the HVAC units of the building according to the scheduling prepared by the strategy planning module (for example, the scheduling plan can be a series of on off commands for the HVAC units). In turn, The activity of HVACs influences on the climate measured by the indoor sensors.

4   Heat Transfer Model

In this section, building on top of the works done for the thermodynamic modelingof houses, [1, 25, 27] we describe a simplified thermodynamic model adapted for the heat transfer inside a building. The ultimate goal of our work in the model, is to demonstrate how a physics-based model can be adjusted applying statistic correction, such that to adapt to different kind of buildings.

4.1   Assumptions

Temperature regulation of buildings is a complex system. A complete theoretical approach is impractical and sometimes can be of difficult realization, as it implies a deep and specific knowledge of building structures and its status. Therefore, here we rely on the following set of assumptions:

1. Air in the zone is fully mixed. Temperature distribution is uniform and the dynamics can be expressed in a lump capacity model
2. Effect of each internal wall is the same
3. The floor and roof have no effect on the zone temperature;
4. The density of the air is assumed to be constant and is not influenced by changing the temperature and humidity ratio of the area.

4.2   Model

The general view of the rate of the heat change in time can be described as:

$$\frac{dQ}{dt} = A(T, t) + B(t) \tag{1}$$

where the A function describes the loss of the heat and B describes the actual heating generation. The heating function B depends only on time, while the heat loss depends also on the temperature. Since $dQ = C * dT$ we can rewrite the formula above for the temperature change:

$$\frac{dT}{dt} = \frac{1}{C}(A(T, t) + B(t)) \tag{2}$$

where C is the thermal capacity. We consider the following types of heat transfer (get/loss):

1. Conduction: we consider the heat transfer between two contacted media with different temperatures, like the indoor and outside walls;
2. Additional heating sources: we consider additional heating/cooling devices impacting on the final temperature in the area;
3. Radiation: we consider the solar radiation that goes through the windows of the area;

Conduction. Let us consider a room where the heat transfer is done via the outside and inside walls (Fig. 2). In our first simplified approximation we consider all the additional heat/cool sources are switched off.

To simplify the model we assume that the temperature inside the room is higher then outside. The heat flow between two regions with different temperature can be described as:

$$\frac{dq}{dt} = -\sum_i h_i A_i (T_1(t) - T_0(t)) - \sum_j h_j A_j (T_1(t) - T_2(t)) \tag{3}$$

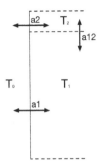

Fig. 2. Modeling the heat transfer with two kind of media

where A_i is the i surface of the heat transfer with h_i heat transmittance coefficient. Here, we do one more important assumption: we consider the medium to have a good thermal conductivity and so to be at uniform temperature. Hence, we can apply the definition of heat capacity C as the relation $C = \frac{dQ}{dT}$ that can be differentiated with regards to time $\frac{dQ}{dt} = C\frac{dT}{dt}$. We can rewrite the Eq. 3 as:

$$\frac{dT}{dt} = -\sum_i \frac{h_i A_i}{C_1}(T_1(t) - T_0(t)) - \sum_j \frac{h_j A_j}{C_2}(T_1(t) - T_2(t)) \qquad (4)$$

where C is the heat capacity of the medium. Since we have two media with heat exchange with outside, the Eq. 4 becomes a system of differential equations:

$$\frac{dT_1}{dt} = -\alpha_1(T_1(t) - T_0(t)) - \alpha_{12}(T_1(t) - T_2(t))$$
$$\frac{dT_2}{dt} = -\alpha_2(T_1(t) - T_0(t)) + \alpha_{12}(T_1(t) - T_2(t)) \qquad (5)$$
$$\alpha_1 = \sum_i \frac{h_i A_i}{C_1}, \alpha_2 = \sum_k \frac{h_k A_k}{C_2}, \alpha_{12} = \sum_j \frac{h_j A_j}{C_{12}}$$

Hence as we can see from the last equation the computation of the heat capacity parameters of the building is not trivial and requires precise knowledge about the construction characteristics, which is not always possible. Therefore, we propose to estimate the α_1, α_2 and α_{12} parameters based on the data received from the outside and inside sensors via a likelihood function (see Sect. 4.3).

Heating Generator. Now let us introduce the heating element into our model. Let us consider to have a power generated device, like an A/C system that is heating/cooling additionally the room environment with power $P(t)$. In the first approximation we consider the simplest case and we do not take into account the movement of the air in the room and the position of the device:

$$\frac{dQ}{dt} = -\alpha(T(t) - T_0(t)) + \eta P(t) \qquad (6)$$

where η is a coefficient of the source efficiency. Following the same logic described above we can derive the temperature changing of the room with the A/C system impact as:

$$\frac{dT}{dt} = -\alpha(T(t) - T_0(t)) + \beta P(t)$$

$$\beta = \frac{\eta}{C}$$

(7)

were the parameter β is supposed to be estimated via the model adjustment (see Sect. 4.3).

Effect of Sunlight. The impact of solar radiation on the room heating can be derived as [18]:

$$\frac{dQ_{solar}}{dt} = A\lambda P_{incident}(t)$$

(8)

where A is the surface area of the window, $P_{incident}(t)$ is the power per unit area delivered by an incident flux of photons and λ is the coefficient of efficiency and depends on the window characteristics.

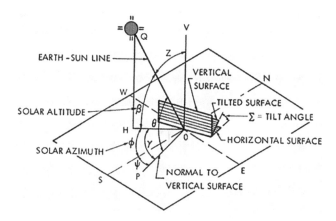

Fig. 3. Geometry of solar position according to the south-faced vertical window (source: [16])

In order to evaluate an incident flux the position of sun in time according to the window in the room has to be considered (Fig. 3). The position of the sun usually is described by two angles: (i) the solar altitude β and the solar azimuth ϕ. Both these angles define the intensity of the solar flux on the surface during the day in function of the time:

$$P_{incident}(t) = P_{average}cos\beta(t)sin\phi(t)s(t)$$

(9)

where $s(t)$ is a discrete function that takes 0 or 1 value and indicates the presence of sunshine at the time.

Hence the resulting differential equation will be:

$$\frac{dT}{dt} = -\alpha(T(t) - T_0(t)) + \beta P(t) + \gamma cos\beta(t)sin\phi(t)s(t)$$

$$\gamma = \frac{\lambda A P_{average}}{C}$$

(10)

where γ has to be adjusted (Sect. 4.3).

4.3   Adjust the Model to Real Data

The model is fed with the data observed by a weather station and the indoor sensors. In other words, we are proposing an approximated model and determine the free parameters of this model, that are α, β and γ parameters in the model described in Sect. 4.2. Based on the collected weather and indoor temperature data (T_i) we infer plausible values for these parameters. In order to tune the parameters of the model we apply the maximum likelihood estimation method.

We assume that the distribution of the values for free parameters in our model follows the Gaussian distribution. In order to explain the approach in the following a free parameter of the model we call θ. We describe the possible noise in the distribution of our parameter θ with a Gaussian function:

$$P(T_i|\theta) = \frac{1}{\sigma\sqrt{2\pi}} \exp(-\frac{(T_{model}(\theta) - T_i)^2}{2\sigma^2})$$

(11)

Hence we can write the likelihood function L:

$$L(T^n|\theta) = \prod_{i=1}^{n} P(x_i|\theta)$$

(12)

or the same but expressed in logarithmic way:

$$\ln L(T^n|\theta) = \sum_{i=1}^{n} \ln P(x_i|\theta)$$

(13)

Hence the most likelihood estimation for the parameter θ can be expressed as:

$$\theta_{mle} = \arg \max(\ln L(T^n|\theta))$$

(14)

Finally, to find the maximum of the likelihood function and to actually estimate the most probable θ parameters for the given (observed) data we have to take the derivative from the likelihood function and equate it to zero.

5   HVAC Scheduling Strategies

In our architecture, the decision block is aimed at choosing the optimal HVAC scheduling that allows to satisfy a given temperature threshold, while minimizing the economical expenses. The scheduling policy is computed according to the predicted indoor temperature $T(t)$, the threshold indoor climate requirements and the energy price plan provided by the energy company, There are several possible template scheduling decisions for the HVAC scheduling that can be divided into three categories (Fig. 4) [10]:

- *Stand by*: In this scenario the system runs HVAC with a constant power in order to keep the current temperature all the time. For example, in order to have a given comfortable temperature in the morning of the next working day, the system can decide do not switch off the HVAC for the night period but instead to leave it to work in order to support the current indoor climate conditions.
- *Basic*: In this scenario the control of HVAC can decide to alternate the switch on and off periods in order to satisfy the desired inside climate conditions [8].
- *Conventional*: This strategy mixes together the previously described ones. The system can keep the HVAC constantly working with reduced power and when it is needed to increase the power in order to reach the target conditions.

In order to optimize the costs of the service, the decision about the strategy should be done based on the energy price plan of the provider. This problem can be formulated as optimization problem, and there are multiple solutions available in the state of the art [11,24,26].

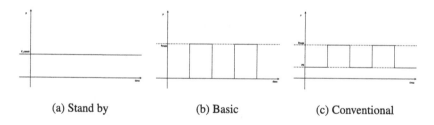

| (a) Stand by | (b) Basic | (c) Conventional |

Fig. 4. Strategies for the HVAC power regulation

6   Conclusion

This paper describes a conceptual architecture for HVAC scheduling in a smart building. The core contribution is the unification of various state of the art approaches and solutions in different fields into a comprehensive architecture.

Among the various components, We detailed a heat transfer model that aims to predict the indoor climate over time according to weather forecast and current temperature values. An accurate heat transfer model has to be adapted to the structural composition of building, to ease the deployment of an HVAC system when applying a regulation system to multiple buildings (or different areas of the same building). Therefore, we advocate the utilization of a model that has the property to automatically adapt to the characteristics of the building. We provided a proof-of-concept heat transfer model for temperature that yield this property.

The architecture presented in the paper does not provide a full coverage of all details of a complex HVAC scheduling system; However, it represents a step toward the identification of its major components, with the ultimate objective of cost saving and energy consumption reduction.

References

1. Achterbosch, G., de Jong, P., Krist-Spit, C., van der Meulen, S., Verberne, J.: The development of a comvenient thermal dynamic building model. Energy Build. **8**(3), 183–196 (1985)
2. Afroz, Z., Shafiullah, G., Urmee, T., Higgins, G.: Modeling techniques used in building HVAC control systems: a review. Renew. Sustain. Energy Rev. **83**, 64–84 (2018)
3. Auffenberg, F., Snow, S., Stein, S., Rogers, A.: A comfort-based approach to smart heating and air conditioning. ACM Trans. Intell. Syst. Technol. **9**(3), 28:1–28:20 (2017)
4. Di Corso, E., Cerquitelli, T., Apiletti, D.: Metatech: meteorological data analysis for thermal energy characterization by means of self-learning transparent models. Energies **11**(6), 1336 (2018)
5. Drgoňa, J., Picard, D., Kvasnica, M., Helsen, L.: Approximate model predictive building control via machine learning. Appl. Energy **218**, 199–216 (2018)
6. Du, Y.F., Jiang, L., Duan, C., Li, Y.Z., Smith, J.S.: Energy consumption scheduling of HVAC considering weather forecast error through the distributionally robust approach. IEEE Trans. Ind. Inform. **14**(3), 846–857 (2018)
7. Erdinc, O., Tascikaraoglu, A., Paterakis, N.G., Eren, Y., Catalão, J.P.S.: End-user comfort oriented day-ahead planning for responsive residential HVAC demand aggregation considering weather forecasts. IEEE Trans. Smart Grid **8**(1), 362–372 (2017)
8. Escrivá-Escrivá, G., Segura-Heras, I., Alcázar-Ortega, M.: Application of an energy management and control system to assess the potential of different control strategies in HVAC systems. Energy Build. **42**(11), 2258–2267 (2010)
9. Hakimi, S.M.: A novel intelligent control of HVAC system in smart microgrid. J. Electr. Syst. Inf. Technol. **4**(2), 299–309 (2017)
10. Haniff, M.F., Selamat, H., Yusof, R., Buyamin, S., Ismail, F.S.: Review of HVAC scheduling techniques for buildings towards energy-efficient and cost-effective operations. Renew. Sustain. Energy Rev. **27**, 94–103 (2013)
11. Jo, H.C., Kim, S., Joo, S.K.: Smart heating and air conditioning scheduling method incorporating customer convenience for home energy management system. IEEE Trans. Consum. Electron. **59**(2), 316–322 (2013)

12. Kavalionak, H., et al.: A prediction-based distributed tracking protocol for video surveillance. In: 2017 IEEE 14th International Conference on Networking, Sensing and Control (ICNSC), pp. 140–145, May 2017
13. Kavalionak, H., Gennaro, C., Amato, G., Meghini, C.: Dice: a distributed protocol for camera-aided video surveillance. In: 2015 IEEE International Conference on Computer and Information Technology; Ubiquitous Computing and Communications; Dependable, Autonomic and Secure Computing; Pervasive Intelligence and Computing, pp. 477–484, October 2015
14. Killian, M., Kozek, M.: Ten questions concerning model predictive control for energy efficient buildings. Build. Environ. **105**, 403–412 (2016)
15. Klepeis, N.E., et al.: The national human activity pattern survey (nhaps): a resource for assessing exposure to environmental pollutants. J. Exposure Sci. Environ. Epidemiol. **11**(3), 231 (2001)
16. Kudav, G., Panta, Y., Yatsco, M.: Design and testing of wind deflectors for roof-mounted solar panels. WIT Trans. Eng. Sci. **74**, 15–27 (2012)
17. Marche, C., Nitti, M., Pilloni, V.: Energy efficiency in smart building: a comfort aware approach based on social internet of things. In: 2017 Global Internet of Things Summit (GIoTS), pp. 1–6, June 2017
18. McQuiston, F.C., Parker, J.D.: Heating, ventilating, and air conditioning: analysis and design (1982)
19. Oldewurtel, F., et al.: Use of model predictive control and weather forecasts for energy efficient building climate control. Energy Build. **45**, 15–27 (2012)
20. Pérez-Lombard, L., Ortiz, J., Pout, C.: A review on buildings energy consumption information. Energy Build. **40**(3), 394–398 (2008)
21. Reynolds, J., Rezgui, Y., Kwan, A., Piriou, S.: A zone-level, building energy optimisation combining an artificial neural network, a genetic algorithm, and model predictive control. Energy **151**, 729–739 (2018)
22. Shaikh, P.H., Nor, N.B.M., Nallagownden, P., Elamvazuthi, I., Ibrahim, T.: A review on optimized control systems for building energy and comfort management of smart sustainable buildings. Renew. Sustain. Energy Rev. **34**, 409–429 (2014)
23. Teeter, J., Chow, M.Y.: Application of functional link neural network to hvac thermal dynamic system identification. IEEE Trans. Ind. Electron. **45**(1), 170–176 (1998)
24. Tyukov, A., Shcherbakov, M., Sokolov, A., Brebels, A., Al-Gunaid, M.: Supervisory model predictive on/off control of HVAC systems. In: 2017 8th International Conference on Information, Intelligence, Systems Applications (IISA), pp. 1–7 August 2017
25. Walsh, P., Delsante, A.: Calculation of the thermal behaviour of multi-zone buildings. Energy Build. **5**(4), 231–242 (1983)
26. West, S.R., Ward, J.K., Wall, J.: Trial results from a model predictive control and optimisation system for commercial building HVAC. Energy Build. **72**, 271–279 (2014)
27. Wyrzykowski, M., et al.: Thermal Properties, pp. 47–67. Springer, Cham (2019)
28. Xiang Zhao, H., Magoulès, F.: A review on the prediction of building energy consumption. Renew. Sustain. Energy Rev. **16**(6), 3586–3592 (2012)

Work in Progress Papers - General Track

Statistical Model Based Cloud Resource Management

Mitalee Sarker[(⊠)] [iD] and Stefan Wesner [iD]

Institute of Information Resource Management, Ulm University, Ulm, Germany
{mitalee.sarker,stefan.wesner}@uni-ulm.de

Abstract. In this paper, we present a statistical model based VM placement approach for Cloud infrastructures. The model is motivated by the fact that more and more resource demanding applications are deployed in Cloud Infrastructures and in particular, communication data rate and latency bound applications are suffering from common placement algorithms. Based on a requirements analysis from the use cases of the Cloud-Perfect Project and the bwCloud production infrastructure, the need for a network-aware VM placement is motivated. The solution approach is inspired from the data source modelling applied for statistical multiplexer components in ATM networks. For each VM deployed in the Cloud Infrastructure, a probability for data rate distributions is derived from the collected data traces and the overall network resource consumption is estimated by overlaying the individual data rate probability distributions. The second part of the paper outlines a possible integration into a cloud infrastructure using OpenStack as an example. The paper concludes with a discussion on the stability of the model and initial results derived from collected data traces along with the future work.

Keywords: Cloud Data Centre · Network · VM placement

1 Introduction

In recent years, the adoption of Cloud Infrastructure has not only increased in numbers but more and more resource demanding and business critical applications are migrated from dedicated infrastructure towards shared Cloud based solutions. Examples include, but are not limited to, High Performance Computing (HPC) simulations or Data intensive computing (DIC) applications. These applications require large amounts of compute and storage resources and are often executed as distributed or even parallel applications involving a significant amount of low-latency communication among the hosted Virtual Machines (VM) [4,6,11]. To cope with the increasing number of cloud applications, data centres are expanding at a high rate by deploying hundreds of thousands of servers and other necessary equipment [5].

A major Cloud benefit is the ability to react in a flexible manner to changing resource demands. The ability to deploy additional virtual servers in a short

© Springer Nature Switzerland AG 2019
M. Coppola et al. (Eds.): GECON 2018, LNCS 11113, pp. 107–115, 2019.
https://doi.org/10.1007/978-3-030-13342-9_9

time and also release them if no longer needed is often referred as elasticity [7]. Besides adding additional resources, the distribution of virtual servers across the physical infrastructure is not static. For placing a Virtual Machine, a common approach is to define a set of pre-defined flavors with pre-defined number of virtual CPUs and virtual memory capacity as static parameters. When a user has chosen a specific flavor, the deployment algorithm searches for the first fitting or randomly selected host that meets the demand in terms of free memory and virtual CPUs below the maximum allowed overbooking factor [9]. Other parameters such as network load or usage pattern are commonly not considered as decision parameter as it is considered to be sufficiently addressed by overprovisioning of bandwidth[1] in the network [16], whereas additional information such as latency requirements or underpinning switch topology is neglected.

Each virtual server is competing with other already deployed virtual servers on the same physical host. As the resources are shared such as, the same network component/channel, CPU, memory etc., the overall application performance delivered by virtual servers distributed across the Cloud Data Centre is affected similarly. For example, placing the components of a latency-sensitive application in the distant physical hosts, causes delay and affects the performance of that application [10]. As stated in [10], network equipments such as switches, Network Interface Cards (NIC), transmission links etc., also induce latency which in turn triggers performance degradation of many cloud applications. The placement of virtual servers based on static parameters impacts the Cloud operator by making no optimal use of the offered resources potentially increasing operational costs. From users' perspective, the sub-optimal placement impacts the performance of their cloud hosted applications but, similarly also the quality of service for other users and vice versa.

2   Problem Statement

As the resource utilisation and behaviour of the virtual servers are potentially changing very dynamically, it is important to find an appropriate balance between calculating the optimal distribution of resources across all virtual servers by considering also external factors (e.g. cost/energy optimisation) and the time needed to find the configuration and implement the changed configuration. To better understand the problem, the following critical questions need to be taken into account: how to model the communication behaviour of VMs or the set of applications hosted within the VM? How to collect sufficient data from the network traffic and network device to derive accurate models for making fast placement and migration decisions? How to solve the 'black box' problem where the VM is unaware of the physical infrastructure and the system knows only about hardware but nothing about what is happening inside the VM?

The challenge that needs to be addressed is to achieve an initial placement decision that is not only based on static parameters but also on resource demands

---

[1] While more appropriate wording would be data rate we use the established term bandwidth in this document.

that vary over time. Despite the fact that the placement decision is local by its nature (placement ultimately is realised on a specific physical server), it requires a system wide perspective because in cloud systems, decisions for adding or removing new virtual server are taken continuously. The major challenges to be addressed are

1. As decision parameters (e.g. network bandwidth requirement, CPU load, ...) change over time faster than optimisation algorithms can find a new virtual server distribution and much faster than an implementation of a new distribution by migrating VMs, time-series based optimisations are not promising [12,14].
2. Overbooking physical resources is a common approach to address time varying resource demands. The assumption taken here is that the average load stays most of the time (e.g. 95th percentile) below the available resources and no significant performance degradation is experienced. This assumption is only valid if there is no correlation between the hosted virtual servers and high load is not co-scheduled.
3. Another approach to cope with resource demanding applications running inside virtual servers is to either place them in an exclusive region with no or low overbooking or apply certain distribution approaches such as placing only one such server on a physical server and distributing the heavy workload across the system.

3   Related Work

A set of network-aware VM placement and migration schemes have been investigated. A system called "Oktopus" is described in [1], which deploys virtual networks and uses an allocation algorithm for placing tenant's VMs in the physical machines. The algorithm has two versions; cluster allocation algorithm for data-intensive applications and oversubscribed cluster allocation algorithm for applications with components. The system uses rate-limiting for enforcing bandwidth at VM, which doesn't consider the dynamic behaviour of the applications at run-time and hence may cause performance degradation. As discussed in [15], the Peer VMs Aggregation (PVA) algorithm determines the communication pattern of the VMs and places the mutual communicative VMs in the same server to decrease the network traffic and increase the energy savings. The approach is rather re-active and they did not inspect the dynamic change in the network traffic load. VM migration overhead was also overlooked. A two-tier VM placement algorithm called Cluster-and-Cut has been presented in [8] considering the traffic patterns and the data centre network architecture. For VM placement, they only considers network resources with respect to cost optimisation. Moreover, the performance constraints of virtual switches used in the data centre network architecture can deteriorate the overall system performance [11].

The aforementioned solution approaches mainly lack pro-active action as they consider the run-time behaviour of the VMs as well as their initial resource demand which may change during runtime. Considering these shortcomings, this paper is targeting to implement a framework called 'Allocation Optimiser' for intelligent placement and migration decision of VMs in a distributed Cloud environment such as Cloud Data Centre and WAN with respect to network resource consumption and energy and operating cost optimisation. The placement and migration decisions will be based on the analysis of historical communication traffic traces combined with real-time monitored traffic of the VMs deployed in a cloud infrastructure and also the performance characteristics of the switch capabilities and topology. The triggering point for VM migration will be the overload at the network interfaces and network resource failure.

4   Solution Approach

In order to address the challenge of an elaborated placement decision, the following approach is proposed:

– The time varying parameters of a virtual server are modelled as discrete states with associated probability to occur. For example, in order to address the network bandwidth requirements, the observed data rate over a time period is analysed and the probability for a virtual server to send/receive within a certain range is calculated. The resulting model is a discrete probability distribution function. This is following the model used within the Asynchronous Transfer Model statistical multiplexing where traffic sources have been modelled in a similar way.
– Furthermore, the decision if a new virtual server still fits on a physical server can now be derived by overlaying the probability distribution functions. Based on this assumption the probability or, overbooking a resource type can be calculated from the combined distribution function and placement decisions can now be taken based on the upper boundary that is allowed for overbooking.

The assumption taken for this model is that the communication behaviour of the VM, or more precisely the set of applications within the VM can be modelled as a set of discrete data rate states that occur with a rather stable probability. If the communication behaviour is from an observer viewpoint completely erratic (e.g. is based on user requests that do not show any recurring behaviour) this approach would not work. This obviously depends directly from the nature of the application. Considering VMs that do Video Stream rendering and delivery the communication behaviour would be clearly predictable whereas for user or device triggered actions this might not be the case. As of now we therefore concentrate on HPC and DIC applications considered to have rather stable operation modes over time. The functionalities of the framework is shown in Algorithm 1.

4.1   Integration with OpenStack

The framework mainly consists of 2 components, Data Provider and Calculator. Figure 1 depicts the overall procedure and interactions.

Algorithm 1. A network-aware VM placement and migration framework

---

1: Get the current mapping between the VMs and the physical servers;

2: Calculate throughput from monitored Tx and Rx data rate for each running VM for a certain period of time;

3: Calculate probability distribution model for each VM;

4: Store the probability models of all VMs in a database;

5: Calculate per server overlay model from the VMs which are running inside it;

6: Store the overlay models of all physical servers in a database;

if new VM deployment request arrives **then**

    execute the allocation algorithm to produce the optimal candidate hostlist;

else

    periodically update the models;

end if;

7: END

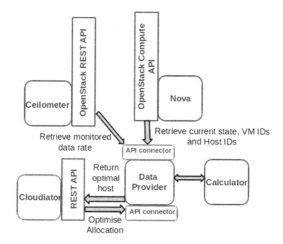

Fig. 1. Integration with OpenStack and Cloudiator tool

Data Provider. This component receives new VM deployment requests from the Cloudiator tool over a REST API. It uses the OpenStack Compute API and the Nova in order to assess the current allocation of running VMs on corresponding physical servers. After getting the VM and server IDs, the component again uses the OpenStack REST API to get the measured time-series data rate values of all implemented VMs from a shared database of a Cloud Monitoring tool such as Ceilometer. For now, in our calculations we consider the measured data for the VMs over the last 24 h. Finally, the Data Provider forwards the list of candidate servers for a new VM back to the Cloudiator tool.

Calculator. This component uses the data rate values which are monitored for a specific amount of time such as 1 day, as an input to estimate the overbooking of bandwidth capacity of the physical servers with respect to the deployment of a new VM and based on that, it produces an optimal candidate server list for the new VM. At first, it calculates histograms by distributing the data rate values over a set of discrete data rate states. Probability distribution models of the data rate for each running VM are then determined by using the histograms. Afterwards, it produces the combined probability distribution model of the data rate for each physical server by overlaying the probabilities of all occurrences on each data rate state for the total number of running VMs per server.

After receiving a request for deploying a new VM from the Cloudiator tool, the Calculator component determines the type of the new VM from it's meta-data and it's related data rate probability distribution model. By overlaying new VM's probability distribution model with the one of each physical server, the tool determines the overbooking probabilities of the bandwidth resource for each server and a list of candidate servers by optimising the energy and operational cost of the Cloud infrastructure. The optimal server list is then sent to the Cloudiator tool via Data Provider component.

After getting the candidate server list, the Cloudiator tool initiates the deployment procedure of the VM. More details of the deployment process can be found in [2,3].

5   Mathematical Representation of the Models

5.1   Probability Distribution Models

After analysing the monitored network data traces, histograms are created by sampling the data rate values onto some specific data rate states such as 10 kbit/s, 8 Mbit/s, 5 Gbit/s etc., for each VM. From the histograms, the probability of the data rate occurrences on the corresponding data rate states for all VMs are calculated by using a simple probability formula [13].

5.2   Overlay Probability Distribution Models

Let's consider a physical server has a virtual machine, VM_i and the total number of Virtual Machine in the server is n.

The Virtual Machine VM_i has now the data rate states as follows:

$$S_{1_{VM_i}}, S_{2_{VM_i}}, ..., S_{N_{VM_i}} \tag{1}$$

The data rate states have the corresponding probabilities:

$$P_{1_{VM_i}}, P_{2_{VM_i}}, ..., P_{N_{VM_i}} \tag{2}$$

Without limiting the model, by setting all other probabilities or states $= 0$ we can assume:

$$N := max\{\text{number of data rate states} VM_1, ..., \text{number of data rate states} VM_n\} \tag{3}$$

Let's assume, a given overlay data rate state, b

$$b = S_{K_{1VM_1}} + S_{K_{2VM_2}} + ... + S_{K_{nVM_n}}, \text{and } [K_i \in \{1, ..., N\}] \tag{4}$$

To simplify the notation, we can write this as,

$$b = b_{K_1 K_2 ... K_n} \tag{5}$$

Then the corresponding probability of the state, b will be

$$P(b_{K_1 K_2 ... K_n}) = \prod_{i=1}^{n} P_{K_{iVM_i}} \tag{6}$$

Let, B be the set of all possible combination of data rate states realising the overlay data rate state, b.

$$B = \{b_{K_1 K_2 ... K_n} | b_{K_1 K_2 ... K_n} = b\} \tag{7}$$

Then, the probabilities of the corresponding overlay data rate states in B will be

$$P(B) = \sum_{b_{K_1} b_{K_2} ... b_{K_n} \in B} \prod_{i=1}^{n} P_{K_{iVM_i}} \tag{8}$$

6   Initial Results

Initial results have been obtained from the monitored data rate values of a set of Virtual Machines running inside bwCloud operational infrastructure. The virtual machines are running a Computational Fluid Dynamics (CFD) application from an user called Nuberisim.

Figure 2 represents the Histograms which have been calculated from the monitored data rate values over 24 h for a Virtual Machine called Nuberisim-worker-01. The X-axis represents the data rate states and the Y-axis shows the occurrences of the data rate values. The size of each data rate state is 10000 bit/s.

Figure 3 depicts the overlayed data rate states and the corresponding probabilities for an hour for two Virtual Machines running inside a physical server.

6.1   Discussion on the Stability of the Discrete Probability Distribution Models

As the Virtual Machine is profiled based on it's network resource usage behaviour, it is essential to determine how stable is the probability distribution model. The stability can be determined by calculating the deviation among the probability values from daily, weekly bi-weekly and monthly data rate probability distribution models of the same running Virtual Machine, where a specific limit of deviation must be selected to define the stability. However, the models can only be valid if they are sufficiently steady and durable with respect to time variance, that means the models should not be updated frequently. Furthermore, the stability of the VM profiles should be evaluated with respect to a set of Virtual Machine instance.

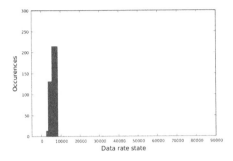

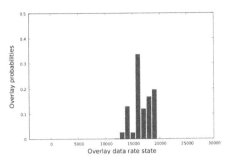

Fig. 2. Histogram for the data rate over 24 h

Fig. 3. Overlay data rate states with corresponding probability distribution for two Virtual Machines

7   Conclusion and Outlook

In this paper, the initial results showed that using simple probability distribution theory, it is possible to estimate the network bandwidth usage of the physical servers which will lead to find an optimal allocation for a new VM to be placed in the Cloud Data Centre. The next steps would be to determine more accurate probability distribution models by using statistical approach such as Hidden Markov Model. Currently the probability distribution models are being calculated based on an average data rate of the VMs for a certain period of time. For developing more definite models, the actual data rate shall be calculated from the inter-arrival time of the packets. Furthermore, in order to determine the limitation of the proposed framework with respect to it's scalability and performance, it needs to be evaluated within a simulation environment including the data centre where different scenarios with varying load distribution and application combinations should be applied. The statistical model is currently determined for estimating network resource usage, but it can also be applied to other resource types such as CPU, Memory.

Acknowledgement. The research leading to these results has received funding from the EC's Framework Programme HORIZON 2020 under grant agreement number 732258 (CloudPerfect). We thank our colleagues from Nuberisim who provided us valuable input that greatly assisted the research.

References

1. Ballani, H., Costa, P., Karagiannis, T., Rowstron, A.: Towards predictable datacenter networks. In: ACM SIGCOMM Computer Communication Review, vol. 41, pp. 242–253. ACM (2011)
2. Baur, D., Domaschka, J.: Experiences from building a cross-cloud orchestration tool. In: Proceedings of the 3rd Workshop on CrossCloud Infrastructures & Platforms, CrossCloud 2016, pp. 4:1–4:6. ACM, New York (2016). https://doi.org/10. 1145/2904111.2904116

3. Baur, D., Seybold, D., Griesinger, F., Masata, H., Domaschka, J.: A provider-agnostic approach to multi-cloud orchestration using a constraint language. In: 2018 18th IEEE/ACM International Symposium on Cluster, Cloud and Grid Computing (CCGRID), IEEE (2018) (accepted)
4. Ferdaus, M.H., Murshed, M., Calheiros, R.N., Buyya, R.: Network-aware virtual machine placement and migration in cloud data centers. In: Emerging Research in Cloud Distributed Computing Systems, p. 42 (2015)
5. Ghiasi, A., Baca, R.: Overview of largest data centers, May 2014. http://www.ieee802.org/3/bs/public/14_05/ghiasi_3bs_01b_0514.pdf. Accessed 19 Apr 2018
6. Jackson, K.R., et al.: Performance analysis of high performance computing applications on the amazon web services cloud. In: 2010 IEEE Second International Conference on Cloud Computing Technology and Science, pp. 159–168, November 2010. https://doi.org/10.1109/CloudCom.2010.69
7. Mell, P., Grance, T.: The NIST definition of cloud computing recommendations of the national institute of standards and technology. http://csrc.nist.gov/publications/nistpubs/800-145/SP800-145.pdf
8. Meng, X., Pappas, V., Zhang, L.: Improving the scalability of data center networks with traffic-aware virtual machine placement. In: 2010 Proceedings of the IEEE INFOCOM, pp. 1–9. IEEE (2010)
9. OpenStackCommunity: Openstack compute schedulers. https://docs.openstack.org/newton/config-reference/compute/schedulers.html. Accessed 06 June 2018
10. Popescu, D.A., Zilberman, N., Moore, A.W.: Characterizing the impact of network latency on cloud-based applications' performance (2017)
11. Sarker, M., Siersch, J., Wesner, S., Khan, A.: Towards a method integrating virtual switch performance into data centre design (2016)
12. Sheridan, C., Whigham, D., Stewart, C., Domaschka, J., Tsitsipas, A., et al.: Validation and result analysis. Cactos project deliverable d7.4.2, revision 3, Institut für Organisation und Management von Informationssystemen (2017). https://doi.org/10.18725/OPARU-4315, open Access Repositorium der Universität Ulm
13. Soong, T.T.: Fundamentals of Probability and Statistics for Engineers. Wiley, Hoboken (2004)
14. Stier, C., Krach, S., Hauser, C., Tsitsipas, A., Domaschka, J., et al.: Performance evaluation of the cactos toolkit on a small cloud testbed. Cactos project deliverable d5.5, Institut für Organisation und Management von Informationssystemen (2017). https://doi.org/10.18725/OPARU-4311, open Access Repositorium der Universität Ulm
15. Takouna, I., Rojas-Cessa, R., Sachs, K., Meinel, C.: Communication-aware and energy-efficient scheduling for parallel applications in virtualized data centers. In: Proceedings of the 2013 IEEE/ACM 6th International Conference on Utility and Cloud Computing, pp. 251–255. IEEE Computer Society (2013)
16. Tso, F.P., Jouet, S., Pezaros, D.P.: Network and server resource management strategies for data centre infrastructures: a survey. Comput. Netw. **106**, 209–225 (2016). https://doi.org/10.1016/j.comnet.2016.07.002

Network Externalities in Cybersecurity Information Sharing Ecosystems

Zahid Rashid[1(✉)], Umara Noor[2], and Jörn Altmann[1]

[1] Technology Management, Economics and Policy Program,
College of Engineering, Seoul National University, Seoul, South Korea
zahid190@gmail.com, jorn.altmann@acm.org
[2] School of Electrical Engineering and Computer Sciences,
National University of Sciences and Technology, Islamabad, Pakistan
13phdunoor@seecs.edu.pk

Abstract. The utilization of cybersecurity information for improving security posture of an organization resulted in the evolution of cybersecurity information sharing ecosystems. In this study, we consider three stakeholders i.e. cybersecurity solution providers, information providers, and end users, who have different values. Their values depend on interrelationship among them and are also based on several value parameters. We identified six value parameters and analyzed their impacts on the values of stakeholders. A simulation model has been developed using system dynamics to analyze the impact of value parameters on the values of stakeholders. The results show that end users are the main source of value in cybersecurity information sharing ecosystems, implying an effect of demand side economies of scale. The cybersecurity solution and information providers are majorly benefiting from a growing number of end users. The value of end users is mainly affected by quality of services, quality of information and the size of trusted communities.

Keywords: Cybersecurity · Information sharing · Ecosystems · Stakeholders · Value parameters

1 Introduction

Cybersecurity information sharing is a promising approach to proactively take necessary measures against potential security threats. This approach is very effective to confront with different threats such as financially driven cyber-crimes, cyberwar, hacktivism, and terrorism [1]. The availability of accurate information at the right time helps organizations to keep themselves updated about ongoing cyberattacks and potential threats. There are several other benefits such as collective defense, collaboration, reduce damage from cyber-attacks, lower cybersecurity incidents, effective response to cyber threats, increased cyber resilience, and low cost on cyber defense [1, 2]. The cybersecurity information sharing ecosystems consists of three major types of stakeholders: cybersecurity solution providers, information providers, and end users. The solution providers deliver software solutions such as antivirus software, threat intelligence platforms, security information and event management systems and other similar solutions, which

© Springer Nature Switzerland AG 2019
M. Coppola et al. (Eds.): GECON 2018, LNCS 11113, pp. 116–125, 2019.
https://doi.org/10.1007/978-3-030-13342-9_10

are used for collecting and processing of cybersecurity information [6]. The information providers provide cybersecurity information in the ecosystems in the form of premium threat feeds, threat intelligence and customer-specific reports. The end users are using cybersecurity solutions and information for improving their cybersecurity postures. The stakeholders of cybersecurity information sharing ecosystems obtain different values and their value creation is interrelated, which creates a complex value distribution structure in the ecosystems. Therefore, it is important to understand value creation and distribution among stakeholders in the ecosystems. As compared to value creation in service platforms and service-oriented technology [4, 5, 15], it is necessary for business managers in cybersecurity information sharing ecosystems to determine the values they obtain from their offered services or products.

The values of stakeholders in cybersecurity ecosystems are influenced by the network externalities, i.e. direct and indirect network effects, also called demand side economies of scale. Direct network effects can be described as values created from the number of existing users, while values generated by the availability of complementary products or services are indirect network effects [14]. The cybersecurity information sharing ecosystems can be viewed as a two-sided market place [17], as information providers attract end users and solution providers. Similarly, solution providers attract information providers and end users. This case is similar to markets of autos discussed in [16], which can be viewed as two-sided market because auto manufacturers must attract both consumers and expert mechanics. The behavior of positive network effect in software and technology product market has been extensively studied in the literature [4, 5, 14] and recognizes the existence of an additional utility in the presence of vendor's large market share.

The cybersecurity solution providers and information providers are not only relying on comparison of functionalities, features, and prices of their products and services but also on end user networks in the form of trusted communities and timeliness of information, which we consider for analyzing the value creation. The trusted community impacts the values of end users and solution providers, because end users can get valuable tools and tips from there for improving responses to cyber-attacks and increase their cybersecurity level.

There are several value creation frameworks available in the literature related to the e-business, software industry and service platforms [4, 5, 11, 15, 20]. In this paper, we presented a value creation framework for cybersecurity information sharing ecosystems, which can provide a useful tool for relevant stakeholders, policymakers, and business managers. The proposed value creation framework will be helpful in explaining the source of values in the ecosystems and also provide support in decision making related to investments in quality of services, business model design, cybersecurity solutions and information bundling policies and evolution of market structures. In this study, we identified six value parameters: installed base (number of end users), trusted communities, quality of services (QoS), cost, quality of information (QoI), and timeliness of information to explain values of stakeholders in the ecosystems. The value parameters are integrated into three utility functions, representing values generated for respective stakeholders. The relative changes in values of stakeholders have been analyzed by simulation technique using system dynamics methodology. We have used the Vensim software of Ventana Systems for our simulations [19].

The results show that end users are the main source of value in cybersecurity information sharing ecosystems and indicates a positive demand side economies of scale. The cybersecurity solution and information providers are majorly benefiting from a growing number of end-users (i.e., installed base). The value of end users is mainly affected by quality of services, quality of information and the size of trusted communities. The size of trusted communities is directly related to the number of end users. The large size of trusted communities has a positive impact on the value of end users. A low number of end users results in low revenues for solution and information providers that ultimately leads to market failure.

The outline of rest of the paper proceeds as follows: Sect. 2 presents an overview of the stakeholders of cybersecurity information sharing ecosystems. Section 3 describes value parameters that affect the value creation of stakeholders. In Sect. 4, the proposed value creation model and its simulation is presented. The results of the model simulation are discussed in Sect. 5. Finally, Sect. 6 provides concluding comments and directions for future research.

2 Stakeholders Involved in Cyber Security Information Sharing Ecosystems

In cybersecurity information sharing ecosystems, there are three major types of stakeholders: cybersecurity solution providers, cybersecurity information providers, and end users. We will discuss each of them in the following subsection. In this study, we are considering the products of only commercial solution and information provider. This means that end users have to pay in return of using solutions and information.

Cybersecurity Solution Providers: Several types of software solutions are being used for generating, consuming and sharing of cybersecurity information. These solutions include antivirus software, security information and event management platforms, network traffic analysis tools, intrusion monitoring platforms, threat intelligence management platforms, forensics platforms, visualization and reporting platforms etc. [6].

Cybersecurity Information Providers: Information related to cybersecurity threats are produced by several open source and commercial vendors. Antivirus software vendors, IT security companies, computer forensic experts, and penetration testers collect cybersecurity information from different sources and after value addition along with services are sold to other stakeholders in the ecosystems [18]. Different pricing models have been adopted by cybersecurity solution and information providers. In several scenarios, one stakeholder can adopt multiple roles simultaneously in the ecosystems, for instance, solution provider and information provider can be same entity.

End Users: All organizations that are using digital technologies in their routine activities or business processes are potential consumers or end users of cybersecurity information. Critical infrastructures, business enterprises, IT companies, cybersecurity companies, government organizations and research organizations are using cybersecurity information for improving cybersecurity posture of their organizations.

3   Parameters Effecting Value Creation of Stakeholders

This section discusses value parameters that determine values obtained by stakeholders in cybersecurity information sharing ecosystems. In our proposed value creation model, we consider six value parameters based on literature in cybersecurity information sharing area. These value parameters are: quality of service (QoS), quality of information (QoI), installed base, timeliness of information, trusted communities and cost.

Quality of Service (QoS): The QoS measures functional and non-functional capabilities of cybersecurity solutions. It indicates whether functionality and performance of cybersecurity solutions meet the requirements of end users. The literature mentions that cybersecurity solutions provide several features and functionalities mainly including: information collection, correlation, integration, enhancement & contextualization, searching & querying, pattern matching, reports generation, distribution & dissemination, automation and integration with existing systems [6, 7, 9, 10]. The QoS is the most important factor that drive values of both cybersecurity solution providers as well as end users.

Installed Base: The number of active users in cybersecurity information sharing ecosystems represents the installed base. The installed base is the main source of revenue in the ecosystems and it affects values of all stakeholders.

Cost: In cybersecurity information sharing ecosystems, cost is used to represent all types of costs incurred by the stakeholders. The cost has a negative effect on the values obtained by stakeholders. The end users have to incur majorly two types of costs: cost of cybersecurity solution and the cost of cybersecurity information. The end users have to pay for subscription which can be either annual or monthly. The cybersecurity solution and information providers face costs for services offered such as maintenance cost and end users support.

Quality of Information (QoI): QoI is an extremely important factor in cybersecurity information sharing ecosystems to improve the value of end users [7–10]. The QoI of cybersecurity information considers same data quality problems as compared to traditional data sets including accuracy, consistency, completeness, trust, and relevance [9].

Trusted Communities: Trusted communities are important part of cybersecurity information sharing ecosystems. Organizations having same goals of achieving a high level of cybersecurity work together and form relationships of cybersecurity information sharing among each other are referred to as trusted communities. The cybersecurity solutions such as TIPs provide platform to enable community collaboration among common interest entities [12, 13]. If installed base of TIP is large, the community size is also large which positively affect values of end users as well as cybersecurity solution providers.

Timeliness: Timeliness is a measure of how cybersecurity information remains current, valid and allow sufficient time for recipient to take appropriate action. Timeliness of cybersecurity information is very important for cybersecurity decision makers in making real-time decisions. The cybersecurity solution and information providers are responsible for ensuring that up to date information is distributed in a timely manner to the end users [9].

4 Value Creation Model and Simulation

This section describes proposed value creation model consisting of effects of parameters on values obtained by stakeholder, the stakeholder's value representation and simulation settings.

4.1 Effects of Parameters on Values Obtain by Stakeholders

The Fig. 1 represents the proposed model showing effects of value parameters on the values of stakeholders in cybersecurity information sharing ecosystems. The parameters are affecting values of stakeholders in positive or negative manner and are represented by (+) and (−) symbols on arrows heads respectively. The installed base has a positive effect on the values of stakeholders and it attracts more end users [11, 12]. The cybersecurity solution and information providers can gain a competitive advantage by leveraging the network of a large number of end users. In such scenarios, network effects also play an important role in attracting more end users and are considered as the business strategy.

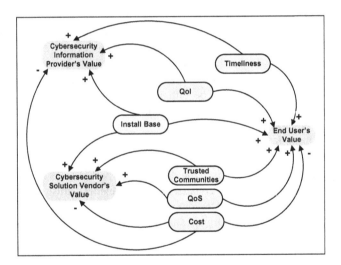

Fig. 1. Effects of value parameters on the values obtained by the stakeholders

In our proposed model we consider that installed base is same for both cybersecurity solution as well as information providers. But in real case, there may be a separate installed base for both types of stakeholders in the ecosystems. The QoS, QoI, timeliness and trusted communities have a positive impact while cost has a negative impact on values of relevant stakeholders. There are majorly two types of cost involved in using cybersecurity information i.e. cost of cybersecurity solutions and cost of cybersecurity information. The cybersecurity information providers are separate entities but in some scenarios, the cybersecurity software solution providers bundle the information from the information provider and sell the complete bundle to the end users.

4.2   Stakeholders Value Representations

This section describes utility functions which are constructed based on the value parameters to quantify values obtained by all stakeholders in the ecosystems. There is a fixed pool of potential users, who may adopt existing cybersecurity solutions and information sources which are added to the installed base. The value parameters are used as inputs to utility functions for calculating utilities (i.e. profit).

End Users Value: Based on Fig. 1, the value of end users is net utility $U_{i,j,a}$ obtained by using functional benefits of utilizing cybersecurity solution and information in the ecosystems. The net utility $U_{i,j,a}$ of end users is defined as follows:

$$
\begin{aligned}
U_{i,j,a}(t) = &u1_{j,a}(QoS(t)) + u2_{i,a}(QoI(t)) + u3_{i,j,a}(IB(t)) \\
&+ u4_{i,a}(T(t)) + u5_{j,a}(TC(t)) - (C_{i,a} + C_{j,a})
\end{aligned}
\tag{1}
$$

Where $U_{i,j,a}$ is a net utility that an end user (a) obtain by utilizing cybersecurity information from an information source (i) and adopting cybersecurity solution (j) at a given time (t). The net utility of end user (a) is the sum of individual positive benefits from all parameters discussed in Sect. 3 minus the respective costs i.e. cost incurred in using cybersecurity solution $(C_{j,a})$ and information $(C_{i,a})$.

Cybersecurity Solution Providers Value: The value of solution provider is the net profit which it gets in return of providing solutions or services in the ecosystems. The value of solution provider is represented by $U_{j,s}$ and is defined as follows:

$$
U_{j,s}(t) = (F_j(t) * IB_j(t)) - C_j(t) - P_{i,k}(t)
\tag{2}
$$

Where $U_{j,s}(t)$ is net value (profit) of solution provider (j) from offering solutions or services at a given time (t). The profit is calculated as the difference between revenue (R_j) and cost (C_j) of cybersecurity solution provider. The revenue of cybersecurity solution provider is calculated as $(F_j * IB_j)$ where (F_j) is an average fee that an end user pays to (j) for obtaining services multiplied with its total installed base (IB_j). The cost (C_j) represents all costs that are incurred for the provision of support to end users, development of new functionalities and maintenance services etc. which increases as size of installed base becomes larger. If solution provider sells its solution by making bundles including information from some other information provider, in this case cybersecurity solution provider have to pay $(P_{i,k})$ to information providers which is deducted from its total revenue.

Cybersecurity Information Providers Value: The value of information provider is net profit which it gets in return for providing cyber threats information in the ecosystems. The value of information provider is represented by $U_{i,k}$ and is defined as follows:

$$
U_{i,k}(t) = (F_i(t) * IB_i(t)) - C_i(t) + P_{j,s}(t)
\tag{3}
$$

Where $U_{i,k}(t)$ is net value (profit) of information provider (i) from offering information services (k) at a given time (t). The net profit is calculated as the difference between total revenue and cost incurred by information provider. The revenue of information provider is calculated as $(F_i * IB_i)$ where (F_i) is an average fee that an end user pays to (i) for obtaining (k) multiplied with its total installed base (IB_i). In case of providing information to solution provider (i.e. bundling of solution and information), the revenue $P_{j,s}(t)$ received from solution provider is also added to net profit of information provider. The cost (C_i) represents all costs that are incurred for supporting end users, new information generation, data storage & processing and maintenance services etc. which increases with size of installed base and new information.

4.3   Simulation Settings

For this study we used the Vensim software [19] for modeling and simulation of value creation dynamics of cybersecurity information sharing ecosystems. The duration of simulation was set to 60 months to see the dynamics of value creation in the ecosystem for relatively longer time period. The cost of solutions and information have varying pricing models based on number of users supported, data size, number of records, data processing and sharing capabilities [3]. The values of Installed base, timeliness, QoS and QoI are normalized in the range between [0, 1], where 1 represent the maximum value and 0 represents the minimum value. In our simulation model, the value parameters are dynamically affecting the values of stakeholders.

The potential end users are those who have not yet adopted any of the cybersecurity solution or information source. The adoption rate of end users vary depending upon the utility of end user $U_{i,j,a}(t)$ and its increase or decrease is directly proportional to value of existing installed base. If $U_{i,j,a}(t)$ becomes zero, there will be no new end users while in case of $U_{i,j,a}(t)$ is greater than zero new end users will be added to installed base depending upon the number of potential end users. The number of potential end users is multiplied by the adoption rate to calculate actual number of new end users. The cases of high and low direct network effects have been simulated and their results are presented in next section. The low network effect is simulated by constant adoption rate of end users while dynamic adoption rate represents the high network effect.

5   Results and Discussions

In our simulation, we run two cases of low and high network effects, to show the dependency of values of end users $U_{i,j,a}(t)$, solution providers $U_{j,s}(t)$ and information providers $U_{i,k}(t)$. The results are shown in Figs. 2 and 3 respectively. The scenario of high network effect in Fig. 3 shows a large increase in the values of stakeholders. In this case, the utilities of all stakeholders change rapidly and the size of the installed base also grow faster because new users are dynamically joining the installed base. In case of low network effects, the new users are joining at a constant rate and the installed base is not expanding at a rapid pace as compared to the scenario of high network effects. Therefore, in low network effects, the values of stakeholders are mainly effected by QoI, QoS, and timeliness instead of the installed base. The small size of the installed base results in smaller trusted communities and, therefore, it does not have much impact on the value of end users. The value of an end user at $t = 0$ is $U_{i,j,a}(t) > 0$, because maintaining positive utility of end users is necessary to successfully keep them in cybersecurity information sharing ecosystems. We observe that the growth of values of end users is more than other two stakeholders because it is affected by the values of both solution and information providers in the ecosystem.

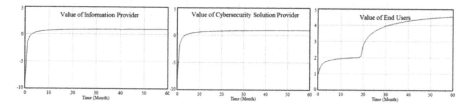

Fig. 2. Constant adoption of end users (low network effect)

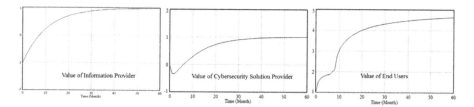

Fig. 3. Dynamic adoption of end users (high network effect)

In both the scenarios of low and high network effects, the value of solution provider $U_{j,s}(t)$ at $t = 0$ is below zero because at the beginning installed base is small and cost of developing and maintaining solutions as well as other costs (such as advertising and other incentives) to attract end users is relatively high as compared to their revenue. From cybersecurity market analysis [3] it has been observed that cost of cybersecurity solution are very high (i.e. from hundreds to thousands of USD) thus end users are very cautious to adopt the new solutions. In case of high network effects, the growing installed base result in large trusted communities which increases the value of solution providers. This increase in value of solution provider has a positive impact on the value of end users. We can observe in Fig. 3 that at $t = 7$ the value of solution providers is moving towards the positive side. Similarly, the value of information provider $U_{i,k}(t)$ is following the same behavior as the solution provider. The value of $U_{i,k}(t)$ at $t = 0$ is on the negative side and moving to positive side at $t = 8$ because initially the install base is small and cost of generating information and cost to attract end users is relatively high as compared to revenue generated. In cybersecurity information sharing ecosystems, values of all stakeholders are interdependent and change in the value of one stakeholder affects the value of other stakeholders. The impact analysis of value parameters shows that the installed base has a mutual positive determinant to the values of all stakeholders. The positive value of end user is important to keep them in the ecosystem, however, sufficiently large values of solution and information providers are also necessary to sustain in the market. In the scenarios discussed above, values of solution and information provider are not very high, which is discussed as negative network effect in the literature [21] of security software market. The value sharing mechanism can be further investigated in which the business models such as bundled solutions (i.e. bundle of cybersecurity solution and information) can be further studied extensively so that solution and information providers sustain their services for longer time periods. The cybersecurity solution providers can benefit more if their solutions can handle information in different formats and standards. These types of solutions are also beneficial for end users as well as for information providers. Similarly, if the cybersecurity information is compatible with most of the cybersecurity solutions than it will add values to all stakeholders in the ecosystem.

6 Conclusions

A value creation model has been presented, which can be used as a tool for evaluating values obtained by stakeholders in cybersecurity information sharing ecosystems. We defined three utility functions that allow integration of value parameters to calculate utility or profit (i.e., values of stakeholders). Six value parameters have been identified from literature: quality of service (QoS), quality of information (QoI), trusted communities, timeliness, installed base, and cost. The value creation dynamics has been evaluated in Vensim software that support the system dynamics simulations. The results of our simulation reveal that the installed base of end users is the main source of value creation in the cybersecurity information sharing ecosystems. The solution and information providers can take benefits of network effects in the ecosystems. It has been observed that positive values of all stakeholders are necessary for stability in the market. The values of all stakeholders are interdependent on each other. The solution and information providers have to devise a strategy to support the values of each other, in order to survive for a long time period and make the market stable. In future, our study will be extended by including more factors such as the detailed market structures, competitive environments, pricing models, and the structure of trusted communities, to establish a fine-grained value creation model in cybersecurity information sharing ecosystems.

References

1. Skopik, F., Settanni, G., Fiedler, R.: A problem shared is a problem halved: a survey on the dimensions of collective cyber defense through security information sharing. Comput. Secur. **60**, 154–176 (2016)
2. Praditya, D., Janssen, M.: Benefits and challenges in information sharing between the public and private sectors. In: Proceedings of 15th European Conference on eGovernment, Portsmouth. UK (2015)
3. Robb, D.: Eight Top Threat Intelligence Companies (2017). https://www.esecurityplanet.com/products/top-threat-intelligence-companies.html. Accessed 01 June 2018

4. Haile, N., Altmann, J.: Value creation in IT service platforms through two-sided network effects. In: Vanmechelen, K., Altmann, J., Rana, O.F. (eds.) GECON 2012. LNCS, vol. 7714, pp. 139–153. Springer, Heidelberg (2012). https://doi.org/10.1007/978-3-642-35194-5_11
5. Haile, N., Altmann, J.: Value creation in software service platforms. Future Gener. Comput. Syst. **55**, 495–509 (2016)
6. Sauerwein, C., Sillaber, C., Mussmann, A., Breu, R.: Threat intelligence sharing platforms: an exploratory study of software vendors and research perspectives. In: Proceedings of 13th International Conference on Wirtschaftsinformatik, Switzerland, pp. 837–851 (2017)
7. Appala, S., Cam-Winget, N., McGrew, D., Verma, J.: An actionable threat intelligence system using a publish-subscribe communications model. In: Proceedings of 2nd Workshop on Information Sharing and Collaborative Security, pp. 61–70. ACM (2015)
8. Fransen, F., Smulders, A., Kerkdijk, R.: Cyber security information exchange to gain insight into the effects of cyber threats and incidents. e & i Elektrotechnik und Informationstechnik **132**(2), 106–112 (2015)
9. Sillaber, C., Sauerwein, C., Mussmann, A., Breu, R.: Data quality challenges and future research directions in threat intelligence sharing practice. In: Proceedings of 2016 Workshop on Information Sharing and Collaborative Security, pp. 65–70. ACM (2016)
10. Brown, S., Gommers, J., Serrano, O.: From cyber security information sharing to threat management. In: Proceedings of 2nd Workshop on Information Sharing and Collaborative Security, pp. 43–49. ACM (2015)
11. Katz, M.L., Shapiro, C.: Technology adoption in the presence of network externalities. J. Polit. Econ. **94**(4), 822–841 (1986)
12. Wagner, C., Dulaunoy, A., Wagener, G., Iklody, A.: Misp: the design and implementation of a collaborative threat intelligence sharing platform. In: Proceedings of 2016 Workshop on Information Sharing and Collaborative Security, pp. 49–56. ACM (2016)
13. ThreatConnect: https://www.threatconnect.com/tag/community/. Accessed 15 June 2018
14. Katz, M.L., Shapiro, C.: Network externalities, competition, and compatibility. Am. Econ. Rev. **75**(3), 424–440 (1985)
15. Demirkan, H., Kauffman, R.J., Vayghan, J.A., Fill, H.G., Karagiannis, D., Maglio, P.P.: Service-oriented technology and management: perspectives on research and practice for the coming decade. Electron. Commer. Res. Appl. **7**(4), 356–376 (2008)
16. Rysman, M.: The economics of two-sided markets. J. Econ. Perspect. **23**(3), 125–143 (2009)
17. Rochet, J.C., Tirole, J.: Two-sided markets: a progress report. RAND J. Econ. **37**(3), 645–667 (2006)
18. Goodwin, C., Nicholas, J.P.: A framework for cybersecurity information sharing and risk reduction. Technical Report, Microsoft (2015)
19. Ventana Systems: Vensim Software Windows Version 7.2 a (2015). http://vensim.com/ Accessed 01 Apr 2018
20. Soto-Acosta, P., Meroño-Cerdan, A.L.: Analyzing e-business value creation from a resource-based perspective. Int. J. Inf. Manage. **28**(1), 49–60 (2008)
21. Dey, D., Lahiri, A., Zhang, G.: Quality competition and market segmentation in the security software market. Mis Q. **38**(2) (2014)

AMFC Tool: Auditing and Monitoring for Cloud Computing

Leandro Pauro[1], Roberta Spolon[1(✉)], Gustavo Bruschi[1],
Aleardo Manacero[2], Renata Lobato[2], and Marcos Cavenaghi[3]

[1] Computing Department, Sao Paulo State University-UNESP,
Bauru, Brazil
leapauro@hotmail.com, roberta.spolon@unesp.br
[2] Computer Science and Statistics Department,
Sao Paulo State University-UNESP, Sao Jose do Rio Preto, Brazil
[3] Humber Institute of Technology & Advanced Learning,
The Business School, Toronto, Canada

Abstract. Cloud Computing has been increasingly incorporated by companies as a cost-effective way to make resources and services continuously available. However, as a consequence of service downtimes at cloud providers, achieving operational reliability and resource availability are still a concern, since they can lead to loss of revenue and customer mistrust. This work presents Apache CloudStack AMFC (Auditing and Monitoring For Cloud Computing), a cloud auditing and monitoring tool aimed to perform the removal of unused data and inconsistencies, improve failure detection (reducing false positive and false negative alerts), and reduce the cost for storing persistent cloud data. All these characteristics are achieved through the synchronization of current state information with persistent orchestration data. The effectiveness of the tool is evidenced through testing on experimental scenarios generated in a controlled test environment. The experiments involved 1,320 administrative routines for virtual machine instances. It was possible to identify and eliminate inconsistencies in the persistent database, allowing a reduction in the storage cost and, consequently, an improvement on database integrity. Overall, the AMFC provided the cloud administrator with more accurate data, enhancing decision-making, allowing a better identification of problems occurring in the cloud environment.

Keywords: Cloud computing · Monitoring · Inconsistency

1 Introduction

Cloud computing is a technology to provide data storage and remote applications that rests on the communication enabled by the Internet and on servers deployed on datacenters. This technology leads to efficient computing environment by consolidating storage, memory, computation and bandwidth [1].

In a cloud environment computational resources are created through virtualization, which is a technique to reserve resources from physical machines encapsulating them in a virtual machine (VM). With this approach it is possible to use, manage and provide services and resources in an optimized way, avoiding that physical resources become idle or overloaded.

© Springer Nature Switzerland AG 2019
M. Coppola et al. (Eds.): GECON 2018, LNCS 11113, pp. 126–134, 2019.
https://doi.org/10.1007/978-3-030-13342-9_11

Failures in cloud systems may occur due to catastrophic events or due to undesired level of service. These failures can be understood as inconsistencies in the services being provided, which involve attending to multiple users, probably with different applications over the time. This service multiplication can lead to inconsistencies such as performance degradation, component failures, or security issues. To maintain services in the levels agreed with users, therefore avoiding inconsistencies, the cloud provider must have efficient mechanisms to prevent or recovery from failures [2, 3].

In this work it is presented AMFC (Auditing and Monitoring for Cloud Computing), a tool for cloud auditing and monitoring aiming the detection and removal of inconsistencies among virtual machines running at the IaaS (Infrastructure as a Service) layer on Apache CloudStack. The remaining text goes through Sect. 2, which covers the related work on auditing and monitoring of cloud environments. The proposed tool is presented in Sect. 3 and Sect. 4 presents the test environment and results. Conclusions and final remarks are presented in Sect. 5.

2 Related Work

Cloud auditing and monitoring can be performed in several ways. Some of the tools presented in the literature use logged data in order to activate certain actions, while others perform dynamic monitoring in order to enable active detection of failures. Along this Section we describe some of these tools and integrated frameworks.

Xu et al. [4] presented a tool that gathers log information coming from VM instances that collect them without interruption. The distributed data can be collected and stored in a single central log by the use of tools like Redis, Logstash, ElasticSearch and Kibana.

Another work presented by Xu et al. [5] uses the SVM statistical learning algorithm (Support Vector Machine) through an API interface in the AWS CloudWatch monitoring tool. This determines which events should be disabled, avoiding that the administrator receives false positive alarms (or does not receive a false negative alarm), or even the occurrence of a flood of different channel alarms on the same event.

Saleh et al. [6] use the concept of Complex Event Processing (CEP) to detect an event or determine the event pattern that has or has not occurred, to coordinate the response action time, and to discover complex patterns among multiple event data streams. Through CEP the authors developed a framework that performs the automated monitoring of the gathered metrics (e.g., CPU utilization, memory usage and disk operations), allowing the management and adjust of resources in real time.

FlexACMS is a framework proposed by Carvalho et al. [7] that performs the integration of different monitoring tools in a cloud environment. This is useful since there is no monitoring tool that integrates all areas of the cloud, and that also satisfies all administrator requirements, the integration requires that administrators manually configure monitoring solutions or develop scripts to automate this task.

MonPaaS is a platform designed by Calero et al. [8], that integrates the monitoring solution Nagios[1] to the orchestrator OpenStack. MonPaaS automatically configures

---

[1] Monitoring tools such as Zabbix (www.zabibix.com), Nagios (www.nagios.org) and Ganglia (ganglia.sourceforge.net/) are used in cloud environments.

Nagios when new cloud slices (a set of resources) are created on the orchestrator. To configure Nagios, MonPaaS execute REST (Representational State Transfer) calls to NConf (Enterprise Nagios Configurator) webservice, a management interface for Nagios. It also relies on DNX (Distributed Nagios Executor) to create Nagios clusters, where one master Nagios server controls several slave Nagios servers to accomplish scalability. MonPaaS also creates a Monitoring Virtual machine (MVM) for each user on the cloud to group all monitoring services of that user.

RMCM (Runtime Model for Cloud Monitoring), presented by Montes et al. [9], uses instrumentation techniques to obtain direct measurements in the application, and interceptors to acquire information about requests that are being processed.

Ceilometer [10] monitors billing, benchmarking, scalability and statistical variables on OpenStack orchestrator. It offers, by the integration and development of a specific API, a set of mechanisms that can be added as alarms and event configurations.

Li et al. [11] proposes a method, based in the Principal Component Analysis (PCA), which optimizes the PCA-DP algorithm used for processing hyperspectral remote sensing on cloud parallel architectures. It uses Apache Hadoop and MapReduce to reduce hyperspectral data dimension and increase its efficiency.

Another open source tool aiming performance monitoring is CloudSurf, proposed by Persico et al. [12]. It measures the performance of the cloud network by traffic capture and analysis.

CloudMonatt performs continuous monitoring of the VMs placed in an OpenStack cloud [13]. It uses the Trusted Platform Module (TPM) for binary attestation in order to execute the monitoring over clients, servers, orchestrator and the hypervisor.

Wang et al. [14] propose a method based in the Canonical Correlation Analysis (CCA) to automate the diagnosis of cloud failures. The main idea is to use CCA to model the correlations between workloads and metrics related to application performance/resource utilization in a specific access behavior pattern.

Failures can be indicated by changes in the virtual machine instances, as performed by DeltaSherlock framework [15].

Du and Li [16] propose the framework ATOM (Efficient Tracking, Monitoring and Orchestration of Cloud Resources) to provide online automated monitoring, orchestration and resource usage monitoring of large scale IaaS clouds.

Another tool concerned with security issues is CloudMon, introduced by Weng et al. [17] for Xen hypervisors. It performs dynamic monitoring to assure security at runtime on kernel operating systems running on a guest VM.

3   AMFC – Auditing and Monitoring for Cloud Computing

The tool developed in this work enables automatic detection of service inconsistencies in a IaaS cloud. It uses Apache's Cloudmonkey for data acquisition, and MySQL Workbench to integrate the cloud orchestrator with the collected data's DBMS. The tool was implemented using Python. AMFC's architecture acts as an intermediate component between the orchestrator and the hosts and VMs. This placement decouples data acquisition from monitoring analysis, allowing the application of the best procedures on each one.

This structure also avoids the need for log files to record the monitored actions. Since the storage of user and administrator's data in a common base is a security problem because an attacker could have access to cloud's administrative information if succeeded [18], the fact that AMFC does not store monitoring data at all is a visible advantage. AMFC performs the monitoring considering that the cloud offers IaaS services. On the IaaS layer we have the set of virtualized components available to consumers as a third-party service. This architecture involves message exchanges among the three major parts, that is the cloud administrator, a VM instance and the persistent database.

The messages are the core of AMFC's operation, which can be understood as a set of steps, starting with data acquisition and going to the identification of inconsistencies, as follows:

Step 1 – Data acquisition. Data acquisition is accomplished by eight different maintenance routines on each instance of virtual machine, being controlled in the orchestrator's console. The information obtained from these routines is stored on a persistent data base for later, possibly offline, analysis.

Step 2 – Monitoring and analysis of information. Monitoring and analysis checks for inconsistent data that can lead to false negative (events that fail to generate a legitimate alert) and false positive (events that generate illegitimate alerts). The analysis includes validation on the synchronization of persistent data stored on the orchestrator's DBMS with information gathered from the cloud environment.

Step 3 – Selecting reaction procedures. The procedures are defined based on the maintenance routines used for data acquisition. This choice follows the techniques adopted on other orchestrators, executing different actions if the event refers to new instances (planned) or to modifications or updates in instance configurations (unplanned).

Step 4 – Monitoring inconsistencies. In the event that some inconsistency is found from the reaction procedures, it is necessary to confirm its existence. This is done by checking if the synchronized information coming from the persistent database matches the one coming from the instrumented cloud. Data inconsistencies appear when an administrator manually changes an aspect of a virtual machine (e.g., removing a data volume), and this change has not yet been reflected in the environment.

Step 5 – Calculation of inconsistencies. The detection of inconsistencies should consider that there are some thresholds to trigger a system reaction. To determine if a threshold has been reached or not, AMFC defines metrics that are calculated from the data acquired in the previous steps. They are:

- Metrics for inconsistencies on resource availability:
  - CTE (cloud test environment): defines the amount of virtual resources present in the system just after the environment's configuration, including users, routers and proxies;
  - APD (amount of persistent data): refers to the amount of database records stored on the persistent database (either consistent or not) related to the cloud resources;

- TIR (total inconsistency per resource): the difference between APD and CTE values, as shown in Eq. 1:

$$TIR = APD - CTE \qquad (1)$$

- RERO (routines executed per resource offer): gives the amount of maintenance routines that are executed per resource in the cloud;
- IPMR (inconsistencies per maintenance routine): is the ratio among the detected inconsistencies and the maintenance routines that were executed, providing a measure of system's efficiency. It is determined by Eq. 2:

$$IMPR = \frac{TIR}{RERO} \qquad (2)$$

- Metrics for proportional gain in the persistent data storage:
  - IPDS (instance's persistent data size): total size of the instance that have cloud persistent data, both consistent and inconsistent, and;
  - SIR (storage improvement per routine): the difference between IPDS and TIR, with the following equation:

$$SIR = IPDS - TIR \qquad (3)$$

- SRO (storage resource offer): is the size of storage available for a given resource;
- SDG (storage data gain): measures the percentage of the storage provided by the resources that is saved by the removal of inconsistencies by AMFC.

$$SDG = \frac{SIR}{SRO} * 100 \qquad (4)$$

4 Evaluation and Results

The test environment involved a private and controlled cloud computing system, built using Apache Cloudstack orchestrator [19]. Although private, it included most of the resources needed by users in an IaaS cloud, such as VM management, network-as-service, user management and accounting and open native APIs. It also supports the major hypervisors in use nowadays, including VMware, KVM, XenServer, Xen Cloud Platform (XCP) and Hyper-V.

The monitoring and auditing actions are centered in the execution of eight different types of routines, they are: 1. Create a data volume disk; 2. Attach a data volume disk; 3. Migrate an instance to another host; 4. Restart an instance; 5. Stop an instance; 6. Remove an instance; 7. Recover an instance; 8. Start an instance. For each size of resource offer (small, medium and big) were created ten VM instances. For evaluation it was performed 440 routines per resource over 30 virtual machine instances, totaling 1,320 administrative routines.

The model described was evaluated, providing the results described in the following paragraphs. The evaluation approach measured the number of inconsistencies occurred in each test, aiming to verify a possible correlation with the number of routines attempted. As we can see on Fig. 1, the number of inconsistencies grows with the amount of maintenance routines performed, despite the model's size. It is also possible to infer that the number of inconsistencies has a positive correlation with the number of administrative routines. Equation 2 was used to calculate the number of inconsistencies.

A different analysis involved the discrimination of how many inconsistencies were created by different types of maintenance routines. The results are presented at Fig. 2, where it is possible to see that operations related to creation and attachment of data volumes (virtual disks) and instance migrations generated most of the inconsistencies. An interesting aspect is that starting and restarting instances do not create inconsistencies.

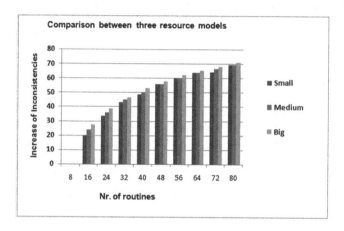

Fig. 1. Comparison between three resource models

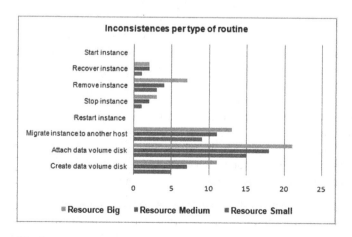

Fig. 2. Number of inconsistencies per type of maintenance routine

Table 1. Space saved by the removal of inconsistencies.

Resource	Disk size (Gbytes)	Space saved (kbytes)	Space saved (in %)
Small	5	30.4	0.00057983
Medium	20	41.6	0.00019836
Big	200	43.2	0.00004120

Although quantifying the number of inconsistencies is an important aspect of cloud management, it is also important to measure the impact of solving such inconsistencies to the system. Among the possible contributions from AMFC's actions one finds saving space in storage by the removal of unused information originated by inconsistencies or data no longer required. Table 1 shows the storage space saved by AMFC for each size of VM instance. Despite the volume saved is rather small considering the size of modern disks, it must be understood that volume is cumulative, i.e., the longer the instances are under operation more useless data will be created. Therefore, the cleanup process must be performed often in order to no degrade the performance.

5 Conclusions

As described in the initial sections, auditing and monitoring a cloud system is very important to achieve system's effectiveness. This motivated the formulation and implementation of AMFC. Particularly, AMFC performs these activities by the synchronization between current cloud information and those stored on persistent storage. To perform these activities AMFC uses Apache CloudStack orchestrator, XenServer hypervisor and Openfiler NAS system. The application of AMFC to a model system provided some insights about its efficiency. From the results presented in the previous section it is possible to draw the following conclusions:

- Inconsistent information due to maintenance routines were automatically removed by AMFC, avoiding flooding administrator with false positive and false negative alerts. This prevents admins from making hasty decisions;
- Accuracy of the failure detection process was improved due to the removal of inconsistent information from the persistent storage. When compared to results in [5], the removal of inconsistencies led to an improved accuracy with AMFC, demanding less time to scan/query;
- AMFC performs auditing and monitoring of virtual machines without creating log files containing monitored information, thus avoiding security problems related with their storage; AMFC is able to remove irrelevant and useless information, reducing uncontrolled redundancy and possible performance degradation;
- AMFC introduced a new approach for auditing and monitoring, where current environment information is related, online, to stored data.

Future developments involve two areas: extending operational capabilities and enhancing functional scope. In the operational side, it is possible to enable alerts to mobile devices, allowing remote management, and to enable live migration of data and

VMs, improving performance. In the functional side it is possible to add the capacity to analyze and identify other types of failures derived from different inconsistencies. In order to conclude, AMFC improves the management of clouds although it still has points for improvement. Its major contribution is using online data to reduce inconsistent management information, avoiding the need for storage and log of possible secure data.

References

1. Alnazir, M.K.A.M., Mustafa, A.B.A.N., Ali, H.A., Yousif, A.A.O.: Performance analysis of Cloud Computing for distributed data center using cloud-sim. In: 2017 International Conference on Communication, Control, Computing and Electronics Engineering, pp. 1–6 (2017)
2. Mdhaffar, A., Halima, R.B., Jmaiel, M., Freisleben, B.: A dynamic complex event processing architecture for cloud monitoring and analysis. In: 2013 IEEE 5th International Conference on Cloud Computing Technology and Science, vol. 2, pp. 270–275 (2013)
3. Suciu, G., Halunga, S., Ochian, A., Suciu, V.: Network management and monitoring for cloud systems. In: Proceedings of the 2014 6th International Conference on Electronics, Computers and Artificial Intelligence (ECAI), pp. 1–4 (2014). https://doi.org/10.1109/ECAI.2014.7090169
4. Xu, X., Zhu, L., Weber, I., Bass, L., Sun, D.: POD-diagnosis: error diagnosis of sporadic operations on cloud applications. In: 2014 Annual IEEE/IFIP International Conference on Dependable Systems and Networks, pp. 252–263 (2014). https://doi.org/10.1109/DSN.2014.94
5. Xu, X., et al.: Crying wolf and meaning it: reducing false alarms in monitoring of sporadic operations through POD-monitor. In: 2015 IEEE/ACM 1st International Workshop on Complex Faults and Failures in Large Software Systems (COUFLESS), pp. 69–75 (2015)
6. Saleh, O., Gropengießer, F., Betz, H., Mandarawi, W., Sattler, K.U.: Monitoring and autoscaling IaaS clouds: a case for complex event processing on data streams. In: IEEE/ACM 6th International Conference on Utility and Cloud Computing, pp. 387–392 (2013)
7. de Carvalho, M.B., Esteves, R.P., da Cunha Rodrigues, G., Granville, L.Z., Tarouco, L.M. R.: A cloud monitoring framework for self-configured monitoring slices based on multiple tools. In: 9th International Conference on Network and Service Management, pp. 180–184 (2013)
8. Calero, J.M.A., Aguado, J.G.: MonPaaS: an adaptive monitoring platformas a service for cloud computing infrastructures and services. IEEE Trans. Serv. Comput. 8, 65–78 (2015). https://doi.org/10.1109/TSC.2014.2302810
9. Montes, J., Sánchez, A., Memishi, B., Pérez, M.S., Antoniu, G.: GMonE: a complete approach to cloud monitoring. Future Gener. Comput. Syst. 29, 2026–2040 (2013)
10. Dongmyoung, B., Bumchul, L.: Analysis of telemetering service in OpenStack. In: 2015 International Conference on Information and Communication Technology Convergence (ICTC), pp. 272–274 (2015). https://doi.org/10.1109/ICTC.2015.7354546
11. Li, Y., Wu, Z., Wei, J., Plaza, A., Li, J., Wei, Z.: Fast principal component analysis for hyperspectral imaging based on cloud computing. In: 2015 IEEE International Geoscience and Remote Sensing Symposium (IGARSS), pp. 513–516 (2015)

12. Persico, V., Montieri, A., Pescapé, A.: CloudSurf: a platform for monitoring public-cloud networks. In: 2016 IEEE 2nd International Forum on Research and Technologies for Society and Industry Leveraging a better tomorrow (RTSI), pp. 1–6 (2016)
13. Zhang, T., Lee, R.B.: Monitoring and attestation of virtual machine security health in cloud computing. IEEE Micro **36**, 28–37 (2016). https://doi.org/10.1109/MM.2016.86
14. Wang, T., Zhang, W., Wei, J., Zhong, H.: Fault detection for cloud computing systems with correlation analysis. In: 2015 IFIP/IEEE International Symposium on Integrated Network Management (IM), pp. 652–658 (2015)
15. Turk, A., et al.: DeltaSherlock: identifying changes in the cloud. In: 2016 IEEE International Conference on Big Data, pp. 763–772. 439 (2016). https://doi.org/10.1109/BigData.2016.7840669
16. Du, M., Li, F.: ATOM: efficient tracking, monitoring, and orchestration of cloud resources. IEEE Trans. Parallel Distrib. Syst. **28**, 2172–2189 (2017)
17. Weng, C., Liu, Q., Li, K., Zou, D.: CloudMon: monitoring virtual machines in clouds. IEEE Trans. Comput. **65**, 3787–3793 (2016). https://doi.org/10.1109/TC.2016.2560809
18. Lin, C.Y., Chang, M.C., Chiu, H.C., Shyu, K.H.: Secure logging framework integrating with cloud database. In: International Carnahan Conference on Security Technology, pp. 13–17 (2015)
19. Cloudstack: Open source cloud computing documentation, 2 April 2018. http://cloudstack.apache.org/

Special Topic Session - IT Service Ecosystems Enabled Through Emerging Digital Technologies

Business Model Characteristics for Local IaaS Providers for Counteracting the Dominance of the Hyperscalers

Sebastian Floerecke$^{(\boxtimes)}$ and Franz Lehner

Chair of Information Systems II, University of Passau, Passau, Germany
{sebastian.floerecke,franz.lehner}@uni-passau.de

Abstract. The Infrastructure as a Service (IaaS) market is dominated by only a few globally acting hyperscalers. The rest consists of a multitude of smaller providers whose IaaS services are restricted to one country or region. As basic IaaS services have become a commodity, the price has turned into the most important decision criterion for customers. For this reason, the central concern of IaaS providers is to achieve economies of scale. However, because of their marginal size, the locally operating IaaS providers are unable to compete in this situation. Accordingly, a growing market consolidation among the local IaaS providers can be expected within the next years. To compete with the further increase in dominance of the hyperscalers, this paper investigates business model characteristics applying to local IaaS providers. The hypotheses were derived from 21 expert interviews with representatives from 17 cloud providers. Due to the exploratory character of this study, the research approach followed the guidelines of the grounded theory method.

Keywords: Cloud computing · Infrastructure as a Service (IaaS) · Local IaaS providers · Business model · Cloud computing ecosystem · Success factors · Grounded theory

1 Introduction

According to a current study of Gartner [1], the market for Infrastructure as a Service (IaaS) is dominated by five globally acting hyperscalers: Amazon Web Services, Microsoft, Alibaba, Google and Rackspace. Amazon Web Services as the leading IaaS provider controls about 44% of the sector [1]. Apart from the hyperscalers, a large number of locally operating, mostly small- and medium sized, providers offer IaaS services, too [2]. These providers exclusively offer their services within one country or region. As already foreseen by Böhm, Koleva, Leimeister, Riedl and Krcmar [3] in 2010, the basic IaaS service model has become a commodity in the meantime. Accordingly, there remain only a few opportunities for IaaS providers to differentiate from one another by their business models. Due to this high degree of homogeneity of the IaaS services among the various providers, the price has become the most important decision criterion for customers. The central issue of IaaS providers is consequently to obtain economies of scale. Only this way, IaaS services can be delivered at comparatively low costs. For local IaaS providers it is impossible to keep pace with this

© Springer Nature Switzerland AG 2019
M. Coppola et al. (Eds.): GECON 2018, LNCS 11113, pp. 137–150, 2019.
https://doi.org/10.1007/978-3-030-13342-9_12

intensive price competition. The hyperscalers are well aware of their position of power and have been continuously pushing down the prices aiming to kick their local competitors out of the market. As a consequence, a growing market consolidation among the local IaaS providers can be observed for a certain period of time. To give an example, United Internet recently acquired Profit Bricks, a medium-sized IaaS provider concentrating on the German market.

To prevent further company acquisitions and thus, cluster building among the local IaaS providers, it is mandatory for local IaaS providers to design and implement differing business models. To the best of the authors' knowledge, business model characteristics influencing the success of local IaaS providers have been, however, neglected in the literature so far. Beyond this background, this paper addresses the following research question: **What are business model characteristics for local IaaS providers to successfully differ from the hyperscalers in order to ensure their long-term competitiveness within the cloud computing ecosystem?**

The paper proposes eight hypotheses on differing business model characteristics for local IaaS providers. The hypotheses were derived from 21 expert interviews with representatives from 17 cloud providers. To maximize insights, experts who work for cloud providers characterized by different experience, size, geographic coverage, target markets and served industries were interviewed. Due to the exploratory character of this study, the research approach followed the fundamental guidelines of the grounded theory method [4].

2 Related Work

Cloud computing represents a new IT operations model that has radically changed the way IT resources are produced, provided and used [5]. The vision that IT services offered from the cloud are commoditized and delivered in a manner similar to traditional utilities such as water, gas and electricity [6] is increasingly becoming a reality. According to the often cited definition of the National Institute of Standards and Technology (NIST), "*[c]loud computing is a model for enabling ubiquitous, convenient, on-demand network access to a shared pool of configurable computing resources (e.g., networks, servers, storage, applications, and services) that can be rapidly provisioned and released with minimal management effort or service provider interaction*" [7]. The five key characteristics of cloud computing services, including on-demand self-service, broad network access, resource pooling, rapid elasticity and measured service, distinguish it from on premise IT solutions [8]. In order to meet the requirements of various customers, four deployment models are available, namely public, private, hybrid and community clouds. These deployment models differ in their degree of operational isolation regarding access to a specific cloud service and the physical location of the servers [5, 7].

With the introduction of cloud computing, both vendors of traditional IT services and start-up companies were given the opportunity to take up new roles in this emerging market [9]. A role can be understood as a "*set of similar services offered by market players to similar customers*" [3]. This evolution was accompanied by a shift from sequential customer-focused IT value chains to complex network-like business ecosystems [10]. A business ecosystem represents a pertinent scope for systemic innovations, where different interrelated and interdependent companies cooperate to deliver full-scale customer solutions [11]. In order to create a profound understanding of the business ecosystem in the context of cloud computing, several attempts of a formal description have been made [12]. A comprehensive role-based ecosystem model is the **Pa**ssau **C**loud **C**omputing **E**cosystem Model (PaCE Model) [13]. Its core consists of providers of the three cloud computing service layers: Infrastructure as a Service (IaaS), Platform as a Service (PaaS) and Software as a Service (SaaS) [7]. Building on these three interrelated service layers, a multitude of further roles, such as aggregators, integrators and market place operators, has emerged [12]. This paper focuses on the infrastructure provider's role, which offers basic infrastructural resources (compute, storage and network) [13].

Each ecosystem role is related to specific business opportunities for providers. Hence, each role must be instantiated by a different business model [14]. A business model can thus be defined as a detailed specification of how ecosystem roles are realized by individual actors [15, 16]. Apart from the business ecosystem view, a business model is seen as a tool for describing, implementing and evaluating the business logic of a firm [17]. Even though no commonly accepted definition of the term "*business model*" has been established yet, the component-based view dominates the research. Accordingly, a business model is a system comprising a set of constitutive components or partial models and the relationships between them [18]. An agreement related to a specific set of relevant components is, however, missing [19]. Nonetheless, a multitude of cross-industry and industry-specific business model frameworks provide design options for selected components [20]. One comprehensive and widespread cross-industry framework is the Business Model Canvas [21], which includes nine components: key activities, key resources, partner network, value propositions, customer segments, channels, customer relationships, cost structure and revenue streams.

Overall, the research on cloud computing business models is nascent [8, 22]. The only comprehensive cloud computing-specific business model framework so far was proposed by Labes, Erek and Zarnekow [23]: it entails eight categories representing the basic components of a business model, further broken down into design features showing possible design options. Labes, Hanner and Zarnekow [24] compared the business models of selected IT service providers with the framework and identified four common patterns of cloud business models. Apart from that, researchers analyzed the fundamental impacts of the shift from delivering on premise IT applications to cloud services (e.g. [8, 10, 25–27]). In addition, scholars have dealt with the process of transforming an on premise to a cloud business model [28, 29]. Ebel, Bretschneider and Leimeister [30] developed and evaluated a software tool for supporting the business model creation. A literature study of Labes, Erek and Zarnekow [22] shows that several further contributions have dealt with one specific or a small number of business model components, such as the revenue [31] or the resource model [32], whereas a holistic approach remains an exception. Investigating them isolated, however, contradicts the logic of business models as the components are interrelated and interdependent [18, 20].

The business model concept became popular after the burst of the dot-com bubble in 2000 [18]. The reason was that scholars were searching for an explanation why a large number of firms had failed, while others had been successful [20]. Thus, the business model concept has played a central role in explaining a firm's performance and deriving success factors for a considerable time [16, 33]. Rockart [34] defines success factors as *"the limited number of areas in which results, if they are satisfactory, will ensure successful competitive performance for the organization"*. Success factors are by definition applicable to all companies of a specific industry with similar objectives and strategies [35, 36]. In this paper, this is substituted by ecosystem roles having their own business model characteristics. A fundamental distinction can be made between generic success factors, which are valid for all kind of companies, and domain-specific success factors, in this case cloud-specific success factors [24]. Hence, it is difficult to transfer the success factors from adjacent research areas to the cloud computing ecosystem without prior examination [37]. Success factors of cloud providers' business models have been addressed by the following studies: Trenz, Huntgeburth and Veit [38] focused on specific success factors regarding the relationship between providers and consumers in the end consumer market. Labes, Hanner and Zarnekow [24] derived abstract success factors by relating publicly available characteristics of the business model components to a firm's web visibility and profit. However, both studies neglected that the cloud computing ecosystem allows the adoption of more than one role and thus, is characterized by a high degree of heterogeneity [13]. Whereas several studies have examined the SaaS provider's role [37], to the best of the authors' knowledge, the infrastructure provider's role and consequently also local IaaS providers are still missing and therefore addressed in this study.

A literature review by Poulis, Yamin and Poulis [39] shows that there is, independently of the cloud computing context, a large amount of literature available which compares multinationals with domestic companies along several dimensions. However, there is a dearth of research on how local firms can compete with the dominating and globally acting companies [39]. According to Chang and Xu [40], this phenomenon has been typically studied from the perspective of multinational firms. This means that local firms have been mostly seen as passive recipients and not as active competitors in a given market [40].

3   Research Design

Quantitative research methods predominantly allow the verification of already formulated hypotheses. As research on differing business model characteristics for local IaaS providers is nascent, it is necessary to further collect data in order to continue and deepen the investigations. Due to this exploratory and hypotheses generating character, the research approach follows the fundamental guidelines of the grounded theory method [4]. *"The grounded theory approach is a qualitative research method that uses a systematic set of procedures to develop an inductively derived grounded theory about a phenomenon"* [41]. According to Wiesche, Jurisch, Yetton and Krcmar [42], the grounded theory method is, however, not exclusively appropriate to develop a theory.

Also models (definitions of abstract variables and their relationships, formulated as hypotheses) or rich descriptions of new phenomena may be the outcome. The targeted contribution is strongly dependent upon the choice of grounded theory procedures [42]. In line with Wiesche, Jurisch, Yetton and Krcmar [42], a partial portfolio strategy was applied as the objective here is a model in the form of hypotheses.

To reach the goal of deriving hypotheses on differing business model characteristics for local IaaS providers, 21 exploratory expert interviews [43] with representatives from 17 cloud providers had been conducted. The 21 experts stemmed from twelve large and five medium-sized cloud providers, had between three and ten years' experience in the cloud field and held leading positions within their companies (board members, portfolio, product, sales, marketing and IT managers, and senior consultants). The cloud providers are characterized by different experience, size, geographic coverage, number of occupied ecosystem roles, target markets, served industries and assessment of the importance of cloud services compared to on premise solutions.

All interviews were based on a pre-tested interview guide, encompassing semi-structured and open-ended questions. The interview guide (available upon request from the authors) focused on deriving business model characteristics influencing the success of IaaS, PaaS and SaaS providers from different perspectives. These perspectives were taken from the literature on success factors and business models. When conducting the interviews in accordance with the grounded theory approach, the authors posed more detailed questions, depending on the flow of conversation, and thus, expanded the basic version of the interview guide. For this purpose, the laddering technique, which follows a process of digging deeper by asking further questions [44], was applied whenever considered appropriate. The interview guide was not sent to the experts in advance deliberately, as spontaneous responses were desired.

The 21 interview sessions took place from June to November 2017. The interview language was German. Ten interviews were done face-to-face, eleven via telephone. Sturges and Hanrahan [45] have shown that there are no significant differences between a face-to-face and a telephone interview with regard to the quality of the gathered data. The duration of the interviews ranged from 30 to 100 min. In order to facilitate the data analysis, all interviews were recorded with the permission of the participants. Each interview was transcribed and proof read. The aggregated transcripts comprised 182 pages of text. As the participants were guaranteed anonymity, the acquired data was sanitized so that no single person or company can be identified.

In line with the grounded theory method, the data analysis started parallel to the data collection and was guided by constant comparison. The data analysis was performed in two phases according to the recommendations of Corbin and Strauss [46] with the qualitative data analysis software MAXQDA. The first phase consisted of open coding – "[t]he process of breaking down, examining, comparing, conceptualizing, and categorizing data" [41]. The derived codes were discussed among the authors and colleagues of the research department in an iterative manner until common agreement was reached. In the second phase, the axial coding technique – relating codes to each other through a combination of inductive and deductive thinking [41] – was applied. This resulted in eight main codes, which represent the derived hypotheses on differing business model characteristics for local IaaS providers. Overall, the whole data analysis was an iterative process of (re-)coding data, splitting and combining categories, and generating new or dropping existing categories. The research process was continued until theoretical saturation was reached [46]. This was the case, when the answers of the interviewees contained no longer new aspects, so that further data collection would not have provided additional insights.

The most of the eight hypotheses are related to aspects for which a market demand exists, which is, for various reasons, not covered by the hyperscalers. Explanations might be that the hyperscalers (i) ignore the opportunities as these stand against their goal of obtaining economies of scale (**H2**), (ii) obey them to a substantial lesser extent (**H1, H3, H4, H7**) or (iii) have deliberately chosen alternatives (**H8**). Furthermore, two hypotheses are based on partnership opportunities offered by the hyperscalers, which seem to be auspicious for local providers (**H5, H6**). Summarized, the business model characteristics for local IaaS providers differ from the current business models of the hyperscalers and can at least partially be explained by customer demands.

4   Local IaaS Providers and Relevant Business Model Characteristics

In order to get a better understanding of the special situation of local IaaS providers and thus, of the derived hypotheses, a representative example for the analyzed providers is presented briefly: Provider *Alpha* is a medium-sized company employing about 150 people in southern Germany. The company operates two main and two smaller data centers, located in two different cities. Customers are firms of all sizes, primary domiciled in the region, but also from the rest of Germany. The service portfolio consists mainly of traditional IT outsourcing and cloud services. Among the cloud services, all three service models (IaaS, PaaS and SaaS) are supported.

The eight hypotheses on specific business model characteristics for local IaaS providers are presented and explained in detail below. Figure 1 illustrates the match of these business model characteristics with the nine components of the Business Model Canvas [21]. As it can be seen, the hypotheses mainly focus on the value propositions, whereas other business model components, such as revenue streams or customer segments, were not mentioned as differing characteristics. **H5** and **H6** were assigned to both the partner network and the value propositions.

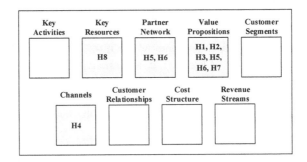

Fig. 1. Mapping the hypotheses with business model components

H1: Offering extensive transition services from on premise infrastructure to IaaS is positively related to local IaaS provider's ability to compete

The interviewed experts stated that medium-sized and large companies have invested quite a lot into on premise infrastructure. Those companies face a considerable challenge when partially moving existing IT applications into the cloud. Only in rare cases, the migration of the systems can be managed without external help. During the transitional period, extensive consulting and customizing support is necessary to get a firm ready for the cloud. The interviewees stressed that the hyperscalers, however, offer such transition services only to a limited extent. Instead, their focus within the market is mostly on firms of all sizes that are already cloud-ready. This gap between clients' demand and the hyperscalers' service portfolio reveals a large opportunity for local IaaS providers. In addition, offering transition services brings further advantages for a provider as he can directly win clients based on his own IaaS service portfolio.

H2: Offering customer-specific adaption of IaaS services is positively related to local IaaS provider's ability to compete

IaaS services from the hyperscalers are characterized by a very high level of standardization. This is the only way to achieve the targeted economies of scale. But, according to the interview partners, some customers have additional requirements that cannot be entirely met by standard services. In this light, it appears promising for local providers to address the discrepancy between the standardized services of the hyperscalers and the specific requirements of certain customers. For this purpose, local IaaS providers have to preserve a certain degree of flexibility within their IaaS service portfolio, even though this is against the basic logic of cloud computing at a first glance. Of course, customization is associated with additional costs, but a willingness to pay can be expected if an added value can be guaranteed. Particularly local IaaS providers have a great advantage as their organization often is more flexible which allows them to respond faster to individual customer demands.

H3: Offering extensive customer support is positively related to local IaaS provider's ability to compete

Receiving extensive customer support for the whole cloud service lifecycle is, according to the interviewees' experience, essential for most customers. Especially for local IaaS providers, who have lower innovative strength and limited sources to react on low prices, customer support can be an option to differentiate against the hyperscalers. A lot of customers appreciate a personal contact and are willing to pay extra for high quality support. Support services include services related to the selection, implementation and operation of cloud services. Local IaaS providers should closely work with their customers as they want to call for help anytime a problem occurs. In contrast, the hyperscalers are often criticized for their unsatisfactory support processes. Offering additional support for hyperscalers' IaaS services therefore seems to be a further option which can complement the own service portfolio.

H4: Offering personal sale instead of self-service sale is positively related to local IaaS provider's ability to compete

On demand self-service is a definitory characteristic of the cloud concept [7]. According to the interview partners, it is associated with possible cost savings as the sales staff can be reduced and standardized contracts can be used. Moreover, the entry barrier of ordering a cloud service and the duration of the process to win a new customer is lowered. For this reason, the self-service option is being enforced by the hyperscalers. However, the experts stated that firms of various sizes differ in their acceptance of self-services: whereas small companies often decide to use the self-service option, medium-sized and large companies commonly prefer personal contact to the provider combined with individual contract negotiations. In general, the self-service variant only makes sense in combination with standardized cloud services when no individual adjustments are needed. In addition, practice shows that several clients have difficulties in using the self-service order process. The main reason for this is a lack of skills on the customers' side. Summarized, it seems to be promising for local IaaS providers to put more emphasis on personal sale and direct interaction with customers.

H5: Offering managed services as an extension of basic IaaS services is positively related to local IaaS provider's ability to compete

Standard IaaS services consist of basic virtual compute, storage and network resources. Besides that, there is, according to the experts, a growing market for managed services. Managed services are IaaS services that entail an extension comprising elements such as update, monitoring or backup services. The main reason for the popularity of managed services is a lack of skills on the customer's side. Especially among firms that formerly were traditional IT outsourcing customers a high demand for managed services can be found. These companies are used to pass the responsibility for the complete IT operations on to the provider. Innovative start-ups, in contrast, often prefer basic IaaS services. Overall, it appears auspicious for local IaaS providers to profit from this growing market for managed services. The managed services can be delivered on the basis of the own as well as the hyperscalers' basic IaaS services. In practice, more and more hyperscalers actively mandate smaller partner firms to take over the managed service part for their IaaS services.

H6: Offering multi-cloud management and reselling of hyperscalers' IaaS services is positively related to local IaaS provider's ability to compete

The interviews showed the growing importance of enabling and offering multi-cloud management. The underlying idea is to offer one's own IaaS service and additionally act as a broker for other providers. This is because of the simultaneous use of IaaS services from various providers by the majority of the medium-sized and large companies: Some SaaS services need to be deployed on the IaaS/PaaS platform of the respective provider. Furthermore, employees sometimes order IaaS services without prior approval by the IT department. And last but not least, customers try to avoid vendor lock-in. For local IaaS providers, this leads to opportunities to benefit from the cooperation. A prerequisite is that local IaaS providers make sure that their IaaS services are compatible with those of the hyperscalers. In addition, providers have to develop and offer a tool to centrally control the various utilized IaaS services and to orchestrate workloads between different clouds. By the means of multi-cloud management, customers will be served by a single point of contact. Offering multi-cloud management is, however, a challenge, mainly due to the lack of uniform standards between the leading IaaS providers. It seems to be reasonable to additionally act as a trusted advisor. This means that the provider tries to select an appropriate IaaS provider for the customers' specific requirements.

H7: Offering private and hybrid cloud deployment models is positively related to local IaaS provider's ability to compete

The demand for private cloud solutions is currently significantly higher than for public clouds. Private clouds are preferred due to data protection, security, availability, regulation and compliance reasons. In addition, a private cloud allows a substantially higher degree of customization which customers often demand. One way to realize a private cloud is a dedicated environment in the provider's data center. Another solution could be to deliver the cloud platform as a bundle consisting of soft- and hardware components which then will be integrated in the customer's data center. The interviewees emphasized that the private cloud has to be necessarily considered together with the public cloud offering. This means that the public cloud complements and expands the private cloud, so that customers can shift workloads between the different systems easily. Clients can thus choose the right time to move from private to public cloud. The interview partners also stressed that the restriction on public cloud services is not recommendable at the moment. The hyperscalers also offer private clouds, but primarily focus on public deployment models. For local IaaS providers private clouds are therefore a must in their cloud service portfolio.

H8: Using an open source IaaS platform is positively related to local IaaS provider's ability to compete

The majority of the hyperscalers utilize a proprietary IaaS platform. This can, however, have a deterrent effect on certain customers. To use an open source IaaS platform instead, e.g. Open-Stack, gives advantages to both clients and providers. This would help to avoid the well-known vendor lock-in. By using an open source IaaS platform, clients can easier switch from one IaaS provider to another, if both providers support the open standard. In addition to that, IaaS providers can save the royalty payments. These savings can be passed on to the clients. Providers who are actively involved in the open source community will benefit from the accumulated know-how. They receive regular updates on improvements and participate in the sharing of experiences and best practices. Finally, open standards are a prerequisite to realize cloud native microservices for SaaS solutions. Of course, small and medium-sized local providers can hardly afford to build an IaaS platform from scratch. Nevertheless, it can make sense to adapt the basic version of an open source IaaS platform. Due to the above-mentioned advantages, an open source platform seems to be an auspicious option for local IaaS providers.

5 Discussion

The results show that IaaS providers who exclusively offer their services within one country or region have clear advantages: they can focus on local regulations and security concerns. IaaS providers addressing a broader or even a global market face the challenge of fulfilling all these various country-specific requirements at the same time. The common approach is to establish a central and uniform cloud platform, hoping that it will meet most customers' demands with only slight country-specific adaptions. Nevertheless, a global IaaS provider usually needs more time to adapt to the rapidly changing market conditions compared to the local providers. Due to their smaller company size, local IaaS providers are much more flexible.

However, the hyperscalers benefit strongly from their company size in another way: they can pass the savings generated through economies of scale and technological progress on to the customers. Local IaaS providers cannot compete with the hyperscalers without adapting their business model because they are unable to keep up with technological innovations in the price-sensitive IaaS market. Therefore, it is mandatory for local IaaS providers to stand out by other features. According to the eight formulated hypotheses, local providers should focus on providing additional services beside their basic IaaS services. Additionally, in cooperation with the hyperscalers, it seems to be promising for local IaaS providers to take over the managed service part for hyperscalers' IaaS services and to act as a broker for hyperscalers by offering multi-cloud management. Moreover, it is necessary to offer private and hybrid clouds together. Finally, the use of an open source IaaS platform is associated with numerous advantages, in particular, the avoidance of a vendor lock-in and the support of cloud native applications.

However, at this point it has to be noted once again that the recommendations are intimately connected with the current business models of the hyperscalers and the existing customer demands. This means in reverse, if the hyperscalers radically modify their business models or customer demands change fundamentally, the propositions for local IaaS providers will also be affected. Figure 2 summarizes the hypotheses on the impact of business model characteristics concerning the competitive strength of local IaaS providers in a model.

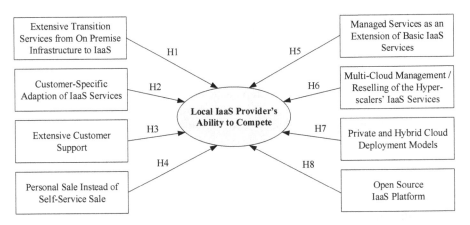

Fig. 2. Hypotheses on differing business model characteristics for local IaaS providers

A major improvement of the situation could be, to additionally take over the role of a PaaS provider. In contrast to IaaS, PaaS offers considerably greater opportunities to generate unique selling proposition and thereby, to differ from other providers. The interviewed experts stated that PaaS is becoming increasingly popular among customers as they can immediately use fully-fledged cloud services. This means, a growing number of providers have ready-to-use PaaS services in their portfolio, which customers formerly had to develop on their own upon an IaaS service. As PaaS often includes elements of machine learning and artificial intelligence, it can provide considerable added value for customers. However, it has been also warned because of the requisites and skills needed for a successful implementation in the PaaS segment. Especially for local providers it is difficult to cope with the innovativeness and speed of the hyperscalers.

It is evident that focusing on the proposed business model characteristics will not transform a local IaaS provider into a global player. The interview partners agreed that this opportunity is no longer given since the hyperscalers are too far ahead. Instead, the business model characteristics should be regarded as orientation help or recommended scope of actions for local IaaS providers to remain competitive. Although some business model characteristics (**H2**, **H3**, **H4**, **H5**) result in higher prices compared to the basic IaaS services from the hyperscalers, a sufficient amount of customers will pay for the significant added value. Nonetheless, the experts predict a shrinking market for local IaaS services. This prediction is mainly based on the assumption that the global players will continue to reduce their prices aiming to kick smaller competitors out of the market. In addition, cloud certificates are gaining importance in the IaaS field and their influence on purchase decisions is expected to increase further in future. Certificates are often demanded within tendering procedures and decision makers will rely on a certified IaaS provider. However, as the procedure of obtaining a certificate is time-consuming and expensive, small and medium-sized local providers are not able to compete in this regard. Because of this, a further growing market consolidation is very likely in the next years. If this happens, the IaaS market is, according to the experts, expected to become subject of governmental regulations, similar to the market for electrical energy. Otherwise, the hyperscalers would use their dominance for arbitrary pricing.

6 Conclusion

This paper addressed the research question of "**What are business model characteristics for local IaaS providers to successfully differ from the hyperscalers in order to ensure their long-term competitiveness within the cloud computing ecosystem?**" Following the fundamental principles of the grounded theory approach, the study's results comprise eight hypotheses on business model characteristics specifically related to local IaaS providers. These hypotheses were derived from 21 exploratory expert interviews with representatives from 17 cloud providers. In detail, local IaaS providers should offer additional services on top of their basic IaaS services. This includes supporting the transition from on premise infrastructure to cloud-based IaaS solutions, but also customer-specific adaption of IaaS services, extensive customer support for the whole cloud service lifecycle, a personal sales contact instead of self-service sale and managed services. Summarized, many customers value personal attention. Moreover, there is the promising option to cooperate with the hyperscalers:

local IaaS providers may to take over the managed service part for hyperscalers and act as a broker for them by offering multi-cloud management. Further recommended actions regard the mix of deployment models, open source platforms and including PaaS in the service portfolio.

Overall, the study provides first insights into business model characteristics which influence the local IaaS provider's ability to compete. Practitioners obtain recommendations and hints that can be useful for improving current business models. The findings, however, have some limitations: First, the geographic scope of interviewed experts was Germany. Second, the hypotheses mainly focus on the value propositions, whereas other business model components, such as revenue streams or customer segments, were not mentioned. They could play a role and future studies, thus, should address these limitations.

Despite of the results achieved, there remains a substantial need for further research: First, the eight business model characteristics are initial hypotheses, which have to be empirically tested. As not all business model characteristics are of equal importance, their relevance might be investigated in a second step. Of course, this exploratory study cannot claim to have identified all possible impact factors for local IaaS providers. Therefore, it is thirdly necessary to research further business model features which contribute to the market position of local IaaS providers.

It will be interesting to watch the evolution of the IaaS market: Firstly, how long can the local IaaS providers withstand the pressure of the hyperscalers and to what extent will the forecasted market consolidation indeed happen? Secondly, will we see changes of the IaaS business models in the light of the growing diffusion of cloud native applications? Furthermore, an exciting question is whether PaaS and SaaS will also become a commodity over the next years.

To conclude, despite the undoubtedly difficult market situation for smaller, local IaaS providers, the authors are quite optimistic that there will always be a niche market for them, if they obey their specific strength which corresponds to the business model characteristics described in this paper.

References

1. van der Meulen, R., Pettey, C.: Gartner Says Worldwide IaaS Public Cloud Services Market Grew 31 Percent in 2016. Gartner (2016)
2. Henkes, A., Heuer, F., Vogt, A., Heinhaus, W., Giering, O., Landrock, H.: Cloud Vendor Benchmark 2016: Cloud Computing Anbieter im Vergleich. Experton Group (2016)
3. Böhm, M., Koleva, G., Leimeister, S., Riedl, C., Krcmar, H.: Towards a generic value network for cloud computing. In: 7th International Workshop on Economics of Grids, Clouds, Systems, and Services, Ischia, Italy (2010)
4. Glaser, B.G., Strauss, A.L.: The Discovery of Grounded Theory: Strategies for Qualitative Research. Aldine, New York (1967)
5. Marston, S., Li, Z., Bandyopadhyay, S., Zhang, J., Ghalsasi, A.: Cloud computing – the business perspective. Decis. Support Syst. 51(1), 176–189 (2011)
6. Buyya, R., Yeo, C.S., Venugopal, S., Broberg, J., Brandic, I.: Cloud computing and emerging IT platforms: vision, hype, and reality for delivering computing as the 5th utility. Future Gener. Comput. Syst. 25(6), 599–616 (2009)

7. Mell, P., Grance, T.: The NIST Definition of Cloud Computing. NIST (2011)
8. Clohessy, T., Acton, T., Morgan, L., Conboy, K.: The times they are a-chaning for ICT service provision: a cloud computing business model perspective. In: European Conference on Information Systems, Istanbul, Turkey (2016)
9. Leimeister, S., Böhm, M., Riedl, C., Krcmar, H.: The business perspective of cloud computing: actors, roles and value networks. In: European Conference on Information Systems, Pretoria, South Africa (2010)
10. Hedman, J., Xiao, X.: Transition to the cloud: a vendor perspective. In: Hawaii International Conference on System Sciences, Manoa, USA (2016)
11. Moore, J.F.: The Death of Competition: Leadership and Strategy in the Age of Business Ecosystem. Harper Business, New York (1996)
12. Floerecke, S., Lehner, F.: A revised model of the cloud computing ecosystem. In: Altmann, J., Silaghi, G.C., Rana, Omer F. (eds.) GECON 2015. LNCS, vol. 9512, pp. 308–321. Springer, Cham (2016). https://doi.org/10.1007/978-3-319-43177-2_21
13. Floerecke, S., Lehner, F.: Cloud computing ecosystem model: refinement and evaluation. In: European Conference on Information Systems, Istanbul, Turkey (2016)
14. Iivari, M.M., Ahokangas, P., Komi, M., Tihinen, M., Valtanen, K.: Toward ecosystemic business models in the context of industrial internet. J. Bus. Model. **4**(2), 42–59 (2016)
15. Tian, C.H., Ray, B.K., Lee, J., Cao, R., Ding, W.: BEAM: a framework for business ecosystem analysis and modeling. IBM Syst. J. **47**(1), 101–114 (2008)
16. Zott, C., Amit, R., Massa, L.: The business model: recent developments and future research. J. Manag. **37**(4), 1019–1042 (2011)
17. Veit, D., et al.: Business models – an information systems research agenda. Bus. Inf. Syst. Eng. **56**(1), 55–64 (2014)
18. Wirtz, B.W., Pistoia, A., Ullrich, S., Göttel, V.: Business models: origin, development and future research perspectives. Long Range Plan. **49**(1), 36–54 (2016)
19. Foss, N.J., Saebi, T.: Business models and business model innovation: between wicked and paradigmatic problems. Long Range Plan. **51**(1), 9–21 (2018)
20. Burkhart, T., Krumeich, J., Werth, D., Loos, P.: Analyzing the business model concept – a comprehensive classification of literature. In: International Conference on Information Systems, Shanghai, China (2011)
21. Osterwalder, A., Pigneur, Y.: Business Model Generation: A Handbook for Visionaries, Game Changers, and Challengers. Wiley, New Jersey (2010)
22. Labes, S., Erek, K., Zarnekow, R.: Literaturübersicht von Geschäftsmodellen in der Cloud. In: Internationale Tagung Wirtschaftsinformatik, Leipzig, Germany (2013)
23. Labes, S., Erek, K., Zarnekow, R.: Common patterns of cloud business models. In: Americas Conference on Information Systems, Chicago, Illionis (2013)
24. Labes, S., Hanner, N., Zarnekow, R.: Successfull business model types of cloud providers. Bus. Inf. Syst. Eng. **59**(4), 223–233 (2017)
25. Boillat, T., Legner, C.: From on-premise software to cloud services: the impact of cloud computing on enterprise software vendors' business models. J. Theor. Appl. Electron. Commer. Res. **8**(3), 39–58 (2013)
26. Morgan, L., Conboy, K.: Value creation in the cloud: understanding business model factors affecting value of cloud computing. In: Americas Conference on Information Systems, Chicago, USA (2013)
27. DaSilva, C.M., Trkman, P., Desouza, K., Lindič, J.: Disruptive technologies: a business model perspective on cloud computing. Technol. Anal. Strateg. Manag. **25**(10), 1161–1173 (2013)

28. Khanagha, S., Volberda, H., Oshri, I.: Business model renewal and ambidexterity: structural alteration and strategy formation process during transition to a cloud business model. R&D Manag. **44**(3), 322–340 (2014)
29. Kranz, J.J., Hanelt, A., Kolbe, L.M.: Understanding the influence of absorptive capacity and ambidexterity on the process of business model change – the case of on-premise and cloud-computing software. Inf. Syst. J. **26**(5), 477–517 (2016)
30. Ebel, P., Bretschneider, U., Leimeister, J.M.: Leveraging virtual business model innovation: a framework for designing business model development tools. Inf. Syst. J. **26**(5), 519–550 (2016)
31. Al-Roomi, M., Al-Ebrahim, S., Buqrais, S., Ahmad, I.: Cloud computing pricing models: a survey. Int. J. Grid Distrib. Comput. **6**(5), 93–106 (2013)
32. Herzfeldt, A., Floerecke, S., Ertl, C., Krcmar, H.: The role of value facilitation regarding cloud service provider profitability in the cloud ecosystem. In: Khosrow-Pour, M. (ed.) Multidisciplinary Approaches to Service-Oriented Engineering, pp. 121–142. IGI Global, Hershey (2018)
33. Lambert, S.C., Davidson, R.A.: Applications of the business model in studies of enterprise success, innovation and classification: an analysis of empirical research from 1996 to 2010. Eur. Manag. J. **31**(6), 668–681 (2013)
34. Rockart, J.F.: Chief executives define their own data needs. Harvard Bus. Rev. **57**(2), 81–93 (1979)
35. Leidecker, J.K., Bruno, A.V.: Identifying and using critical success factors. Long Range Plan. **17**(1), 23–32 (1984)
36. Freund, Y.P.: Critical success factors. Plan. Rev. **16**(4), 20–23 (1988)
37. Floerecke, S.: Success factors of SaaS providers' business models – an exploratory multiple-case study. In: 9th International Conference on Exploring Service Science, Karlsruhe, Germany (2018)
38. Trenz, M., Huntgeburth, J., Veit, D.: How to succeed with cloud services? Business & Information Systems Engineering, pp. 1–14 (2017)
39. Poulis, K., Yamin, M., Poulis, E.: Domestic firms competing with multinational enterprises: the relevance of resource-accessing alliance formations. Int. Bus. Rev. **21**(4), 588–601 (2012)
40. Chang, S.J., Xu, D.: Spillovers and competition among foreign and local firms in China. Strateg. Manag. J. **29**(5), 495–518 (2008)
41. Strauss, A., Corbin, J.M.: Basics of Qualitative Research: Grounded Theory Procedures and Techniques. Sage Publications, Newbury Park (1990)
42. Wiesche, M., Jurisch, M.C., Yetton, P.W., Krcmar, H.: Grounded theory methodology in information systems research. MIS Q. **41**(3), 685–701 (2017)
43. Myers, M.D., Newman, M.: The qualitative interview in IS research: examining the craft. Inf. Organ. **17**(1), 2–26 (2007)
44. Corbridge, C., Rugg, G., Major, N.P., Shadbolt, N.R., Burton, A.M.: Laddering: technique and tool use in knowledge acquisition. Knowl. Acquis. **6**(3), 315–341 (1994)
45. Sturges, J.E., Hanrahan, K.J.: Comparing telephone and face-to-face qualitative interviewing: a research note. Qual. Res. **4**(1), 107–118 (2004)
46. Corbin, J.M., Strauss, A.: Basics of Qualitative Research: Techniques and Procedures for Developing Grounded Theory. SAGE Publications, Thousand Oaks (2008)

Delivering a Systematic Framework for the Selection and Evaluation of Startups

Ece Erdogan and Somayeh Koohborfardhaghighi[✉]

Amsterdam Business School, University of Amsterdam,
Plantage Muidergracht 12, 1018 TV Amsterdam, The Netherlands
ece.erdogan@student.uva.nl,
s.koohborfardhaghighi@uva.nl

Abstract. The literature shows that the failure rate of startups is around 90%. Therefore, it is crucial for investors and financial advisors to be able to spot the 10% which eventually will generate higher return rates and bring in greater revenues. The absence of a general conceptual framework which could assist large corporations and investors in the selection and evaluation of startups is quite visible in the literature. In this research, critical success factors for strategic alliance making between startups and large sized companies are identified and possible selection methods are discussed. Second, based on our findings a conceptual framework is presented for the selection of successful startups. Semistructured interviews are conducted at a large scale financial tech company to evaluate our proposed framework. The results of our expert interviews indicate that all the managers who were involved in the selection process of startups agree on the fact that the team experience and the startup's position within its network are highly related to the success of the startup in the future. Furthermore, characteristics of the lead entrepreneur, competitive advantage of the firm's products and the valuable resources the startup has are also ranked among the criteria which managers look into and have strong influence on their decision making.

Keywords: Strategic alliance · Startups · Digital services · Value co-creation · Expert interviews · Success factors · Service innovation · Exploratory case study

1 Introduction

The literature suggests that the failure rate of the startups is around 90%. Although there may be several reasons behind this rate, including but not limited to lack of financial funds or simply bad management, it is crucial for investors and financial advisors to be able to spot the 10% which eventually generates higher return rates and bring in greater revenues (Krishna et al. 2016). Being able to spot the successful startups is even more important for larger corporations who are willing to take on a partnership with them. This is owing to the fact that when startups enter in to an alliance with corporations, they tend to stay with their partners (due to high switching costs) and their growth eventually contribute to the success of their larger counterparts. Forming strategic alliances has been proven to be very beneficial for the startups too.

© Springer Nature Switzerland AG 2019
M. Coppola et al. (Eds.): GECON 2018, LNCS 11113, pp. 151–159, 2019.
https://doi.org/10.1007/978-3-030-13342-9_13

These strategic alliances with large enterprises may help startups with having strong grounds to build their enterprise on at the very beginning. Having a secure partnership with a large enterprise may also help startups in reducing the impact of a project failure at the early stages (Comi and Eppler 2009, Kinyenje 2016, Baum et al. 2004, Barney 1991, Das and Teng 2000).

Investors are highly interested in determining the common features of the successful startups which are able to bring an innovation into a marketplace. How these startups can be spotted at their early stages is also an important topic concerning large corporations seeking a partnership with them. In order to answer this question, Galloway (2017) suggests that it is essential for a company to create a product which disrupts the market they are operating in. Taking the example of the most successful companies today, he indicates that their common feature was to bring an innovation or operate differently from the incumbent firms which would simply create a disruption. Startups such as Uber and Airbnb also stand as good examples within this context. Uber challenges other means of transport by offering a simpler and cheaper version of transport. Additionally, Airbnb offers a cheaper accommodation where customers can customize their stay according to their preferences and with the help of technology (Galloway 2017). The question then arises as: Is creating a disruptive technology the only common feature of the successful startups?

As there are various literature around the business models that startups have followed to grow as big as they are now, there is a gap in the literature on which criteria can large businesses select the right startups. Therefore, in this research, we aim to address the following research question: What considerations should the large corporations look into for assessing whether or not a startup will be successful in the future? Within the context of value co-creation, big corporations such as Google and Salesforce have devised tools to help firms in selecting which companies they should invest in or form a strategic alliance with. Google's market finder digital services enable large corporations to narrow down their search for a strategic partner. Google's market finder tool sends its clients possible leads and partners that they might be interested in forming a strategic alliance with.

Currently the firms decide on which startups to partner with according to various criteria including but not limited to the startup's growth rate and the initial funding that they may have received. However, the initial comments of the employees of a large enterprise have indicated that, there should be a conclusive list of criteria which would help with the selection of startups as partners. This list would also be beneficial for any size of corporation a strategic alliance will be formed with. Therefore, the contribution of this paper is twofold. First, critical success factors for making strategic alliance between startups and large sized companies will be identified and possible selection methods will be discussed. Second, based on our findings a framework will be presented for the selection of successful startups. In order to answer our research question, semi-structured interviews are conducted at a large scale financial tech company.

The research presented in this study has the following implications. First, the outcome of this research is beneficial for large corporations as it aims to find a roadmap for corporations when they are taking on new partnerships. Secondly, it investigates the importance of the identified criteria within our proposed framework from the point of view of managers who are involved in the process of startup selections. The rest of this paper is organized as follow: In Sect. 2 the relevant literature review on the topic of strategic alliance is presented. In Sect. 3, the proposed framework and an extensive explanation of the research design are delivered. The results of our analysis are presented in Sect. 4. In Sect. 5 we provide the conclusion and discussion over our findings.

2 Literature Review

2.1 Selection Criteria for Selecting Startups

As with the increase of technological focus of the companies and the innovation projects the firms want to take a part in increases, more alliances are being formed with the startups (Comi and Eppler 2009). Startups make it easier for firms to manage their innovation projects and enter in to new markets where technological capacities are a must. Thus, interest in forming alliances with startups have increased recently (Comi and Eppler 2009). According to (Duchesneau and Gartner 1990) there are three characteristics that differentiate a successful startup from the others. Firstly, the characteristics of the lead entrepreneur is an important indicator of their future success. Secondly, the processes that are taken up during the initial growing stages of the startup are highly important. Lastly, the strategic decisions the startups make once they start to grow and become scale ups are also important factors that affect their future success and growth rate. Another factor that differentiates successful startups from other ventures are the long term plans and goals of the entrepreneurs and their vision for the future growth of the startups (Gelderen et al. 2005). Gelderen, additionally emphasizes that the social connections of the lead entrepreneur and their psychological state might also influence their entrepreneurship skills. In the following we discuss how the different entities such as large corporations, venture capitals, SMB's and investors assess the future success of startups.

According to (Comi and Eppler 2009), before entering into a partnership with a startup, large corporations find it important to analyze the intellectual capability of a startup as an indicator of their future success as a partner. This is mainly because it has been found in the literature that existing resources of a venture is an important indication of a suitability of a partner in a strategic alliance. Furthermore, the large corporations find the connections of the founding team of the startup and the financial investments they have acquired at their growth rate as important criteria when selecting them as partners.

In a different research focusing on the selection criteria of the venture capitalists on the selection of startups, it has been found that venture capitalists focus on the early partnerships that the startup might have (Baum and Silverman 2004). The authors show that, venture capitalists take in to consideration any patents that these startups might have acquired as they are an indicator of the technical abilities of the startups. They may specifically look in to the capabilities of the founding team, the intellectual capability of a startup and also the strategic alliances they have. The results of another study indicated that the venture capitalists look into the investment activities, due diligence activities and information that the startups have (Zinecker and Bolf 2015). In a relevant literature around the selection criteria of venture capitalists on startups, it was found out that asking cognitive questions at the selection process might help acquiring more information on startups. This might eventually lead to better understanding of the success rate of startups in the future (Csaszar et al. 2006). Asking cognitive questions to the founders of startups is also essential in case there is not enough information on the startups and if the startups have not provided enough data when approaching the investors and other entities for funding options (Csaszar et al. 2006).

Venture capitalists suggest that it is not possible to predict the success rate of a startup correctly as there are unknown variables that cannot be taken into account. However, analyzing a startup with the aid of asking questions might eliminate risks and help the investors have better understanding of the future success rates of the startup. The authors suggested that these questions should be relevant to the fields of strategy, teams and finance.

In addition to the above criteria, SMBs find the trust and the personal relationships between the SMBs and the startups as one of the most important criteria on the selection process of startups that they enter in to an alliance with (Hoffmann and Schlosser 2001). This is mainly because the trust in the capabilities of the lead entrepreneur and the founding team are important credentials in determining the future success of the startups.

For the investors, having a liaison with a venture capitalist play an important role in shaping the growth of the startups (Baum and Silverman 2004). Therefore, they take into account in which startups the venture capitalists have initially invested in. They also assess the intellectual capability of the startups. Investors also take into account the Initial Public Offering (IPO) stages of the startups when they are assessing their future growth rate. The success rate of the startups especially at the stage of their initial public offerings also depends on three criteria. Firstly, the investments from reputable financial institutions and the venture capitals are important as they provide a sense of trust for the future investors. Additionally, the startups can enjoy further effects of the success of these financial institutions in their next alliances (Chang 2004). Secondly, making strategic alliances with big corporations is an indicator of future growth, as they have access to the important social and technological resources (Chang 2004). These resources will eventually help the startups to grow within their industries. Lastly, it has been proved that these alliances improve the performance of both sides. Therefore, it is stated by Chang that having strategic alliances increases the chances of having a successful IPO process for the technology startups. The author also showed that the amount of strategic alliances that has been formed by the startup and also the reputation of the firms that the startup has formed a partnership with are also important criteria.

3   Proposed Framework and Methodology

Our proposed conceptual framework has been presented in Table 1. As we can see 27 factors have been selected during our literature review. In order to develop our conceptual framework which would yield a list of criteria for the selection of startups, we performed a literature review to determine the right set of factors which influences the selection process. All references related to the presented items are listed in this Table. In order to answer our research question we conduct an exploratory case study at a

Table 1. Our proposed conceptual framework with a conclusive list of criteria for the selection of successful startups.

AUTHOR/YEAR	METHOD	CONTEXT: General	CONTEXT: Venture Capitals	CONTEXT: SMBs	Startup's alliances	Patents	team experience	Strategic Compatibility	Governance mechanism	Reputation of VC	Financials raised by startup	Number of strategic alliances	IPO Events	Cognitive Criteria (Interviews)	Characteristics of Lead Entrepreneur	Processes undertaken during founding	Strategic behaviours after founding	Prior Knowledge on startups	Startup's position within its network	Competitive Advantage of products	Growth rate of the market	Management's familiarity with market	Political factors	Economical factors	Size of the startup	Characteristics of top management team	Relational Factors	History of past strategic alliances	Financials from reputable firms	Cooperative Cultures	Power dependency
Gulati, 1999	Quantitative	x																x	x												
Gulati, 1995	Quantitative	x																x	x								x	x			
Muthoka & Kilika, 2016	Qualitative			x																			x	x							
Schmidt & Ochan, 1977	Quantitative	x																						x	x		x	x			
Comi & Eppler, 2009	Qualitative			x	x	x	x																	x	x	x			x		
Hoffmann & Schlosser, 2001	Qualitative			x				x	x									x	x												
Baum & Silverman, 2004	Quantitative		x		x	x	x	x																					x		x
Brouthers et al, 1995	Qualitative	x				x	x	x							x															x	
Duchesnau & Gartner, 1990	Both	x													x	x	x														
Gelderen et al., 2005	Qualitative	x									x	x			x	x			x							x					
Zinecker & Bolf, 2014	Qualitative		x								x	x	x			x				x	x	x									
Csaszar et al., 2006	Qualitative	x												x						x	x	x									
Chang, 2004	Quantitative		x							x	x	x	x																		
Kinyenje, 2016	Quantitative	x						x																		x					
Das, 2000	Qualitative	x						x																							

financial tech company based in The Netherlands. The company is specifically selected to be in the fin-tech sector as this industry includes various disruptive and innovative projects led by the startups. The interviews will be conducted in one single company in order to get the full view of the firm's partnership strategies among its different departments (Baxter and Jack 2008). The interviewees are selected among different departments of the company in order to have a 360° view on the views of different managers on the selection criteria for their partners and startups. The interviewees were mainly the senior managers of the Product Innovation, the M&A and strategy teams who have managed the partnerships and the strategic alliances of the fin tech company. We interviewed 10 managers within the company and during our interview we investigated the significance of the identified factors in the selection process of startups within the company's routine.

4 Results

The results of our analysis have been presented in Fig. 1. The results indicate that all the managers who were involved in the selection process of startups agree on the fact that the team experience and the startup's position within its network are highly related to the success of the startup in the future. Furthermore, characteristics of the lead entrepreneur, competitive advantage of the firm's products and the valuable resources the startup has are also ranked among the criteria which managers look into and have strong influence in their decision making. As it can be seen almost 70 to 80% of the managers agree that intellectual capital (patents), strategic compatibility, strategic behaviors after founding, growth rate of the market, management's familiarity with the target market, economic factors, relational factors, cooperative cultures, power dependency, reputation of participating venture capitals are also among the interesting features that they might look into during a startup's selection process. The further we move on the list it is unlikely that managers reach an agreement on the identified factors. The participants also ranked the IPO events, history of past strategic alliances, processes undertaken during founding and political factors as the least important factors they take into account when selecting a startup to enter into an alliance. Most of the interviewees find the IPO events less relevant compared to the other criteria which was not in line with the obtained result of Chang (2004). Also half of our interviewees mentioned that the processes undertaken during founding and history of past strategic alliances had no effect on the success of a startup as a partner. The interviewees have indicated that the previous success of startups and their strategic decisions prior to the strategic alliance may not affect their future success. That is because, large enterprises will support them with certain resources which were lacking in their previous partnerships. The interviewees have also mentioned that the decisions the startups take after entering into an alliance are considered to be much more important. The previous literature by (Gulati 1995, Duchesnau and Gartner 1990 and Gelderen 2005) had indicated that these two criteria were important in selecting a partner. However, our obtained results indicate that the level of importance of these criteria might have deteriorated over the years. It may also be an indication that currently the managers focus more on the decisions the firms take after forming an alliance rather than their

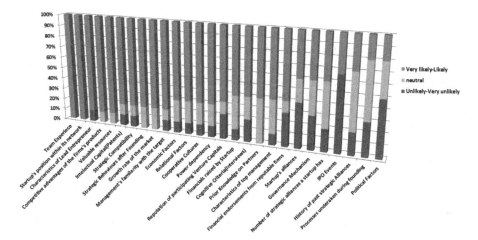

Fig. 1. Respondents were requested to indicate the level of importance of each factor within our proposed framework for the selection of startups.

former strategic decisions. Lastly, 40% of the interviewees do not find it necessarily important to consider prior knowledge on their partners. According to the obtained results they would prefer to partner up with new startups through their connections instead of approaching a new one. Thus, examining or interviewing a new startup's founding team as suggested by Chang (2004), do not provide a lot of benefits to the interviewees. Additionally, another interesting observation was related to the importance of the political factors. Muthoka and Kilika (2016) suggested that such criteria alongside with the economic factors might have an overall effect on the selection of startup. However, the interviewees indicated that they had not considered such factors in their selection process in the past and thus find it irrelevant to consider.

5   Conclusion and Discussion

As with the increase of technological focus of the companies and their innovation projects, more alliances are being formed with the startups. Startups make it easier for firms to manage their innovation projects and to enter in to new markets where technological capacities are a must. The absence of a general conceptual framework in the literature which would help with the selection of startups is quite visible in the literature. Therefore, our main goal in this research was to create a conceptual framework for assessing the selection criteria that should be used by a firm when entering into a strategic alliance with a startup. We performed a literature review to determine the right set of factors which influences this selection process. Next, we conducted an exploratory case study at a financial tech company based in The Netherlands. Semi-structured interviews were conducted at a large scale financial tech company to evaluate our proposed framework. The results of our expert interviews indicated that all the managers who were involved in the selection process of startups agree on the fact that the team experience and the startup's position within its network are highly related to the success of the startup in the future. Furthermore, characteristics of the lead entrepreneur, competitive advantage of the firm's products and the valuable resources the startup has were also ranked among the criteria which managers looked into and had strong influence in their decision making. In addition to the identified factors from the literature review (Presented in Table 1) additional selection criteria were identified by the interviewees which we will elaborate on each of them in the following. (1) Global Scale of the Startups: The interviewees have indicated that it is highly important for a startup to be able to operate globally or have the option to grow globally; (2) Leveraging Diversity for Creative Solutions; (3) Size of the startup; (4) Maturity of the Product of the Startup; (5) Selling Narrative: The selling narrative of the enterprise considers where the product of the startup falls; (6) Startup's Partner Priority: in order to not have conflicts in the future; (7) Timing of the partnership: when does the product of the startup goes in to the market; (8) Feasibility of the startup's product or solution: feasibility requirements or the user experience of the products; (9) The right spending of the funding by the startup: where the startup chooses to spend its initial funding that were received; (10) Being the leader in the target market.

We believe the presented research will be an important contribution to the literature as the proposed conceptual framework can be integrated with existing digital services to evaluate startups systematically. Furthermore, the results of this research can be used by companies who want to participate in the digital service innovation process or delivering social qualities at the system level (Koohborfardhaghighi and Altmann 2015, 2016, 2017). Future research should consider involvement of multiple actors from different industries for the sake of comparison and better assessment of the proposed framework.

References

Barney, J.: Firm resources and sustained competitive advantage. J. Manage. **17**(1), 99–120 (1991)

Baum, J.A., Silverman, B.S.: Picking winners or building them? Alliance, intellectual, and human capital as selection criteria in venture financing and performance of biotechnology startups. J. Bus. Ventur. **19**(3), 411–436 (2004)

Baxter, P., Jack, S.: Qualitative case study methodology: study design and implementation for novice researchers. e Qual. Rep. **13**(4), 544–559 (2008). http://nsuworks.nova.edu/tqr/vol13/iss4/2

Brouthers, K.D., Brouthers, L.E., Wilkinson, T.J.: Strategic alliances: choose your partners. Long Range Plann. **28**(3), 2–25 (1995)

Chang, S.J.: Venture capital financing, strategic alliances, and the initial public offerings of Internet startups. J. Bus. Ventur. **19**(5), 721–741 (2004)

Csaszar, F., Nussbaum, M., Sepulveda, M.: Strategic and cognitive criteria for the selection of startups. Technovation **26**(2), 151–161 (2006)

Comi, A., Eppler, M.J.: Building and managing strategic alliances in technology-driven start-ups. Università della Svizzera italiana (2009)

Das, T., Teng, B.: A resource-based theory of strategic alliances. J. Manage. **26**(1), 31–61 (2000)

Duchesneau, D.A., Gartner, W.B.: A profile of new venture success and failure in an emerging industry. J. Bus. Ventur. **5**(5), 297–312 (1990)

Galloway, S.: The Four: The Hidden DNA of Amazon, Apple, Facebook, and Google. Bantam Press, London (2017)

Gulati, R.: Social structure and alliance formation patterns: a longitudinal analysis. Adm. Sci. Q. 40(4), 619 (1995)

Gulati, R.: Network location and learning: the influence of network resources and firm capabilities on alliance formation. Strateg. Manage. J. 20(5), 397–420 (1999)

Gelderen, M.V., Thurik, R., Bosma, N.: Success and risk factors in the pre-startup phase. Small Bus. Econ. 24(4), 365–380 (2005)

Hoffmann, W.H., Schlosser, R.: Success factors of strategic alliances in small and medium sized enterprises- an empirical survey. Long Range Plann. 34(3), 357–381 (2001)

Kinyenje, M.: Factors That Influence Success of Strategic Alliances Between Mobile Service Providers and Commercial Banks in the Money Transfer Services. University of Nairobi (2016)

Kioko Muthoka, R., Kilika, J.: Towards a theoretical model for strategic alliance, and partner selection among small medium enterprises (SMES): a research agenda. Sci. J. Bus. Manage. 4 (1), 1 (2016)

Krishna, A., Agrawal, A., Choudhary, A.: Predicting the outcome of startups: less failure, more success. In: 2016 IEEE 16th International Conference on Data Mining Workshops (ICDMW) (2016)

Koohborfardhaghighi, S., Altmann, J.: A network formation model for social object networks. In: Zhang, Z., Shen, Z.M., Zhang, J., Zhang, R. (eds.) LISS 2014, pp. 615–625. Springer, Heidelberg (2015). https://doi.org/10.1007/978-3-662-43871-8_89

Koohborfardhaghighi, S., Altmann, J.: How strategic networking impacts the networking outcome: a complex adaptive system approach. In: Proceedings of the 18th Annual International Conference on Electronic Commerce: e-Commerce in Smart connected World, p. 30. ACM (2016)

Koohborfardhaghighi, S., Lee, D.B., Kim, J.: How different connectivity patterns of individuals within an organization can speed up organizational learning. Multimedia Tools Appl. 76(17), 17923–17936 (2017)

Schmidt, S.M., Kochan, T.A.: Interorganizational relationships: patterns and motivations. Adm. Sci. Q. 22(2), 220 (1977). https://doi.org/10.2307/2391957

Zinecker, M., Bolf, D.: Venture capitalists' investment selection criteria in CEE countries and Russia. Verslas: Teorija ir Praktika 16(1), 94–103 (2015)

Service User Perspectives on Delivering Social Innovation: An Implication of the Internet of Things for Business

Olga Maria Plessa and Somayeh Koohborfardhaghighi[(✉)]

Amsterdam Business School, University of Amsterdam,
Plantage Muidergracht 12, 1018 TV Amsterdam, The Netherlands
om.plessa@yahoo.com, s.koohborfardhaghighi@uva.nl

Abstract. Despite the fact that IoT creates many opportunities and drives business growth by increasing the quality and speed of processes, little is known about its potential in delivering social innovation. We aim to deliver a novel framework which has the potential to guide new IoT driven business models in delivering social innovation in line with what service users expect to receive. With the help of an empirical study the significance of different building blocks of the Social Stakeholder Canvas are estimated. Our experimental results show that three building blocks, which are mainly Social Impact (i.e., Privacy and Security), Social Benefits (i.e., Quality of Life), and Scale of Outreach (i.e., Adaptivity and Transparency), are the most influential constructs in delivering social innovation. The findings of this study can provide practitioners with guidelines and reasons to implement new IoT applications and useful insight on how to consider service users insights in delivering social innovations for the society.

Keywords: Internet of Things · Social value · IoT driven business model · Social innovation · Stakeholder Model Canvas · T-test

1 Introduction

Business model describes the way a company creates, delivers and seizes value, or, more simply, how it fulfills its purpose. The business model is essential for the business to survive and grow, always being an integral part of a company. The business models started by being oriented towards output maximization (product oriented) with financial goals as their primary focus. However, traditional enterprises have altered into caring more about innovation and delivering services (Gebauer et al. 2016) to show that business sustainability and shared values are essential to their business plans (Rouse 2013).

The rise of technology, as we observe, cannot leave the business sector and everything it involves unaffected. According to Baden-Fuller and Haefliger (2013), the connection between a business model and the technology is two-way. As a business model changes through recent technologies it should enables and promotes innovation. In a digital society, digital co-creation becomes a business standard. That is to say, by combining business expertise and emerging digital technologies new values should be

© Springer Nature Switzerland AG 2019
M. Coppola et al. (Eds.): GECON 2018, LNCS 11113, pp. 160–168, 2019.
https://doi.org/10.1007/978-3-030-13342-9_14

created for all the actors within a business setting (i.e., customers, stakeholders, people in the society, government etc.) to shape a high-value future. This can be considered as social value creation or delivering social innovation. As social innovation is something that can be referred to as "innovative services aimed at meeting social needs" (Mulgan et al. 2007), it requires the digitalization of business actors' co-creation activities and it aims to realize that value co-creation possibilities are relatively higher as they go beyond geographic constraints (Funaki 2017).

The literature shows that Sociocultural concerns regarding innovation are increasing (Mulgan et al. 2007; Dutton 2014), and this study looks at these concerns through a promising technology, heavily associated to the everyday life, the Internet of Things (IoT). IoT is included by the US National Intelligence Council in the list of six "Disruptive Civil Technologies" with popular demand and future opportunities like contributing to economic development (Li et al. 2015). Despite the fact that IoT creates many opportunities and drives business growth by increasing the quality and speed of processes, little is known about its potential in delivering social innovation. That is to say, we also need to think to what extent does IoT impact society and how it can add to social value. We need to reflect on which social changes are of first priority from the users' point of view and what is the role of IoT in their materialization. The uncertainty surrounding the IoT's future influence makes it significantly harder to question its profound impact on society. Nevertheless, it is undoubtedly a subject worth researching as its analysis can bring new insights regarding the direction that IoT could take towards social innovation. Therefore, the following research question will be addressed: How does a better understanding of service users' priorities improve the contributions of IoT driven business models in delivering social innovation?

The present study focuses on Stakeholder Model Canvas (Joyce and Paquin 2016) for delivering social innovation based on a new IoT driven business model. With a proper literature review on the topic of social innovation we identify relevant components for each building block of Social Stakeholder Canvas. Later with the help of the identified factors we deliver a novel framework, which has the potential to guide new IoT driven business models in delivering social innovation. Our proposed framework is tested with the responses of 327 service users. We perform exploratory factor analysis to test the framework's structure and to capture the significance of the identified factors. The results of this study unveil that some building blocks are significantly more important than others (i.e. Social Impacts, Social Benefits, Scale of Outreach). Items such as Surveillance have negatively scored as they were defined as against the users' values. Our findings can provide practitioners with guidelines and reasons to implement new IoT applications for delivering social innovations in our societies.

The rest of this paper is organized as follow: In Sect. 2 the relevant literature review on the topic of IoT driven business models and social innovations are presented. In Sect. 3, the proposed framework and the research design are delivered. The results of our analysis are presented in Sect. 4. Finally, in Sect. 5 we provide the conclusion and discussion over our findings.

2 Literature Review

2.1 IoT Driven Business Models and Social Innovation

Adopting Internet of Things (IoT) has been a significant worldwide trend for companies. According to Glova et al. (2014), an IoT driven business model can be an important element to an enterprise to unite the technical innovation with the economic perspective. The basic principles of such innovative mindset for designing IoT driven business models are the new nature of products, which should forecast user needs, and outspread product and service personalization (Metallo et al. 2018).

Connected products enable companies to provide better services at a lower cost and minimize the response time. Smart and interconnected items allow revolutionary technologies to open up the scope of new business models for capturing values and create a unique user experience. Moreover, Westerlund et al. (2014) argued that the change of focus from the IoT being primarily a technology platform to viewing it as a business ecosystem drives the evolution of new business perspectives. Authors explained further that "the concept of business model, which is traditionally associated with a single organization's business model, could be replaced with the term 'value design', which is better suited to ecosystems". Finally, the article indicates that challenges of IoT implementation can be overcome by using a business model design tool that considers the holistic identity of the IoT. In their article, Dijkman et al. (2015) chose to use the Business Model Canvas as a framework for the IoT applications because it is based on a meta-analysis of the framework. The Business Model Canvas, which was initially proposed by Alexander Osterwalder, is a template with nine building blocks, which can describe the activities of a company and how they can be changed in order to add value to the firm. This framework designs the business model as a template for the company and is frequently used in business environments.

Although some attempts have been made to explain the changes in the IoT driven business models the issue of delivering social value through this technology has not been addressed yet. Therefore, the need for the creation of business models, where IoT can adapt and simultaneously create benefits for the company and the society has been derived.

Innovation can have several meanings from a simple new way of doing things to a technological breakthrough. This means that it can incorporate any new way to generate value for an organization and for its stakeholders (i.e., customers, suppliers, citizens, government, etc.). As sustainability concerns are on the increase, firms start engaging in sustainability-oriented innovation. Sustainability necessitates progressing from the development of a product to a broader perspective that provides social wealth. Society nowadays expects enterprises to be socially responsible (Herrera 2015). According to Bidmon and Knab (2017) scholars have become increasingly interested in understanding societal transitions, which are large-scale and long-term changes of systems that fulfill societal functions and highlight the importance of business models for achieving systemic changes. The authors even argued that "novel business models have a greater potential to achieve systemic change than technology". Biloslavo et al. (2017) claim that existing business frameworks exclude natural and social aspects and tend to neglect interrelations between economic and non-economic factors and that they need to incorporate the natural environment and future generations in the process of value creation. Sustainable business model archetypes are introduced to describe groupings of mechanisms and solutions that may contribute to building up the business model for sustainability. Bocken et al. (2014) delivered a sustainable business model with a "triple bottom line approach" (economic, environmental and social layer). Joyce and Paquin (2016) introduced a tool to explore a sustainability-oriented business model, called the Triple Layered Business Model Canvas (TLBM). TLBM is meant to incorporate a holistic perspective of the entire business model and it complements the original Business Model Canvas by adding environmental and social layers, which are interconnect "horizontally" (by exploring each value individually) and "vertically" (by integrating value creation across the layers). The social layer of TLBM is developed on a stakeholder's approach, which concerns the groups of individuals that can be influenced by the actions of a company (for example the community, the customer, the employees, etc.). This was decided because the stakeholder perspective is broad and flexible and some of the more significant social impact factors have also been using the same angle.

Particularly, this paper concentrates on IoT driven business models for delivering social value in line with what service users expect to receive. Several scholars have elaborated on social value creation but to the extent of my knowledge, there is no prior research concerning the social impact of IoT. Thus it would be interesting to capture the perspectives of service users in the role of this emerging technology in delivering social values to our society.

3   Proposed Framework and Methodology

In order to develop a well-rounded image over all aspects of delivering social value in a society, a theoretical framework using Stakeholder Model Canvas is created. All of the important social impacts will be examined to understand which could be offered by IoT. Our proposed conceptual framework has been presented in Fig. 2. This Figure presents different "building blocks" of the framework and their relevant factors.

The Stakeholder Model Canvas is chosen because it is constructed by multiple components ("building blocks") that capture all the influences between stakeholders and the organization. Each component of the Stakeholder Model Canvas is called "building block". For example we consider the "Employee" building block as one of the core organizational stakeholder and we aim to capture service use perspectives on what values IoT could provide to the employees.

We identified the relevant factors within each building block through a proper literature review. For instance, Zwetsloot et al. (2013) examined the core values that support health, public safety and wellbeing, and some of them were divided in the adequate building blocks. Furthermore, Kaldaru and Parts (2008) introduced social factors for sustainable economic development and discussed the importance of those determinants in the growth and development. Also according to a report of Grand View Research on IoT in 2016, this technology can impact the security and privacy and the living conditions of the people in the society. It also has the potential to provide them social benefits and improve their quality of life and social well-being. IoT gives the end users feelings of autonomy and control. By heightened sense of control over different situations end users are able to have a better maintenance over their resources, avoiding waste of any kind (i.e., energy, effort, time). Also in the long run companies can find ways to effectively extend their scale of outreach with the usage of IoT technology. This technology can bring transparency and adaptivity to ongoing business operations, which in its own turn improve situational awareness. This technology encourages Individuality of the entities in a business setting. That is to say, by being self-reliant, IoT technology creates a culture which reinforces environmental responsibility. Service users have more involvement in their lives and society which will be a unique experience for them. This technology helps people in achieving freedom and in living without geographical or restrictions. It provides equal opportunities to anyone so that they fulfill their expectations. It is of crucial importance to see the governance in being legally responsible for the misuse of IoT. Due to the continuous observation of the data of people and locations through sensors and devices, a broad range of public safety services can be delivered. However it is government's responsibility to protect the public in terms of privacy and digital crimes. Today's competitive business environment requires the integration of people, process and technology to make the best of institutional capacity and to achieve the optimal results. Through IoT all the system actors have the opportunity to easily communicate with each other.

After defining each building block's factors, we need to capture their significance level. Following Dijkman et al. (2015), the relative importance of the building blocks and the identified factors within them are determined by using the results of a survey. The survey respondents were asked to rate the importance of the identified factors on a 6-point scale, labeled 6 (extremely important), 5 (very important), 4 (moderately important), 3 (slightly important), 2 (not important), 1 (opposed to my values). The survey resulted in 590 responses of which 327 were complete (after removal of records with (st.dev = 0). In order to measure the importance and significance of the building blocks and identified factors, one sample t-tests is performed. One sample t-tests helps us to calculate the variable's mean and determine whether it is statistically different from the building blocks' mean or not.

4 Results

The results of our analysis have been presented in Figs. 1 and 2. Primarily the relative importance of each building block is calculated by comparing the mean of each building block to the mean of all the building blocks that is shown in Fig. 1 (depicted as a horizontal line with the value x = 4.8). The result of running a one sample t-test showed the significance of all building blocks except the End-User and the Social Value building blocks. In this figure the mean of each construct is presentment with a rectangular box while the line presents the range of the data within each construct. It is easy to see that the Social Impact building block with a mean of M = 5.19 is the most important building block in our proposed conceptual model. Another interesting observation is that the Societal Culture building block with a mean of M = 4.474 has obtained the lowest score.

In order to quantify the relative importance of the factors in each building block, we run another sample t-tests for the items within each building block (i.e., mean of each construct as test value). The factors have been placed in the building blocks according to their order of importance (i.e., their calculated means).

The factors above the grey area are the ones that had a significantly more important that the calculated mean of the corresponding building block. Similarly factors below the grey area are the ones that are significantly less important. The grey area demonstrates the factors that have mean score close to the calculated mean of corresponding building block. For example, Social value is the only building block that have 3 items with mean scores very close to the mean of the building block but not enough to make them significantly important. Consequently from the service users perspective, the most important factor in the Local Communities building block is Collaboration, for the Governance building block is Public Safety, for the Employees building block is workplace collaboration, for the Societal Culture building block it is Responsibility, for the End-User building block it is the Efficiency, for the Social Impacts building block it is the Privacy, for the Scale of Outreach building block it is Adaptivity and for the Social Benefits building block it's Quality of life.

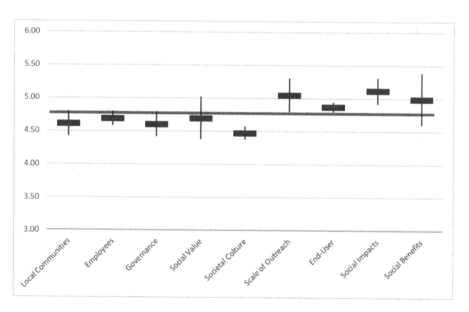

Fig. 1. Relative importance of building blocks.

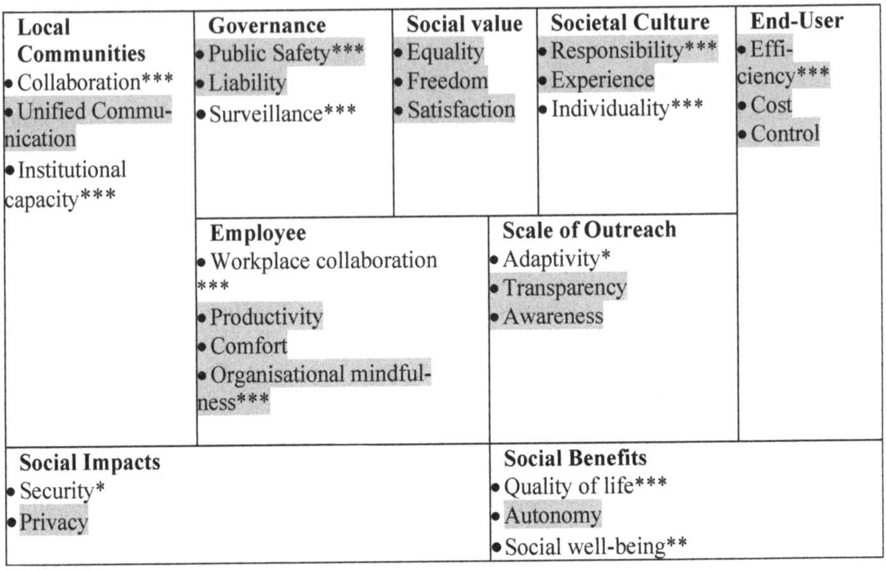

Local Communities • Collaboration*** • Unified Communication • Institutional capacity***	Governance • Public Safety*** • Liability • Surveillance***	Social value • Equality • Freedom • Satisfaction	Societal Culture • Responsibility*** • Experience • Individuality***	End-User • Efficiency*** • Cost • Control
	Employee • Workplace collaboration*** • Productivity • Comfort • Organisational mindfulness***		Scale of Outreach • Adaptivity* • Transparency • Awareness	
Social Impacts • Security* • Privacy		Social Benefits • Quality of life*** • Autonomy • Social well-being**		

Fig. 2. Stakeholder model framework for IoT applications with relative importance of specific types. * < 0.05 significance, ** < 0.02 significance, *** < 0.01 significance

5   Conclusion and Discussion

The literature shows the enormous potential of IoT in delivering innovation as well as sociocultural concerns regarding its development (Mulgan et al. 2007; Dutton 2014). Using Stakeholder Model Canvas, this study aimed to deliver a theoretical framework which cover several aspects of delivering social innovation through this emerging technology in a society. Our experimental results show that three building blocks, which are mainly Social Impact (i.e., Privacy and Security), Social Benefits (i.e., Quality of Life), and Scale of Outreach (i.e., Adaptivity and Transparency), are the most influential constructs in delivering social innovation. Our findings also indicated that our respondents rated the Surveillance factor as "opposed to their values" especially those who have higher education. From the service users perspectives, Public safety, Security, Privacy and Quality of life are rated as extremely important factors. Conceived as very important were Freedom, Transparency, Adaptivity and Autonomy. Security factor appeared to be slightly more valuable for women while Transparency is slightly more important to men.

Innovation is directly linked to value creation for all the actors within a business eco-system. Therefore, by adopting a holistic perspective and considering the holistic identity of the IoT instead of a single organization's business model sustainable development is achievable.

As the extension of tis research we aim to perform exploratory factor analysis, to test the underlying structure of our framework. We also need to improve the assumptions, which we made during the selection of factors within each building blocks of the Stakeholder Canvas Model.

We believe the presented research will be an important contribution to the literature as the proposed conceptual framework can be integrated with existing digital services to deliver social innovation. Furthermore, the results of this research can be used by companies who want to participate in the digital service innovation process (Koohborfardhaghighi and Altmann 2015). Future research should also consider involvement of companies with IoT driven business models for the sake of comparison and better assessment of the proposed framework. That is to say expert interviews should be done in such companies to match the service users expectations with possibilities of products or service offers. Also a critical assessment of the IoT is needed with respect to the social and policy issues raised by its development. A socio-technical perspective also could reflect on diverse human-technology interaction and social needs in the presence of IoT (Shin 2014). The internet of things can also be modeled as a complex adaptive system where the inter-network allowing physical objects to collect and exchange data. Similar to (Koohborfardhaghighi and Altmann 2016a, 2016b, Koohborfardhaghighi et al. 2016, 2017) we are also interested to develop a framework and investigate how the interactions of system's actors (i.e., networking) lead to the emergence of social qualities. The research performed in this thesis has the following implication. As social innovation is something that can be referred to as innovative services aimed at meeting social needs (Mulgan et al. 2007), we expect the outcome of our proposed framework guides new IoT driven business models in delivering social innovation in line with what service users expect to receive.

References

Baden-Fuller, C., Haefliger, S.: Business models and technological innovation. Long Range Plan. **46**(6), 419–426 (2013). https://doi.org/10.1016/j.lrp.2013.08.023

Biloslavo, R., Bagnoli, C., Edgar, D.: An eco-critical perspective on business models: The value triangle as an approach to closing the sustainability gap. J. Clean. Prod. **174**, 746–762 (2017). https://doi.org/10.1016/j.jclepro.2017.10.281

Bidmon, C.M., Knab, S.F.: The three roles of business models in societal transitions: New linkages between business model and transition research. J. Clean. Prod. **178**, 903–916 (2017). https://doi.org/10.1016/j.jclepro.2017.12.198

Bocken, N.M.P., Short, S.W., Rana, P., Evans, S.: A literature and practice review to develop sustainable business model archetypes. J. Clean. Prod. **65**, 42–56 (2014). https://doi.org/10.1016/j.jclepro.2013.11.039

Dijkman, R.M., Sprenkels, B., Peeters, T., Janssen, A.: Business models for the internet of things. Int. J. Inf. Manage. **35**(6), 672–678 (2015). https://doi.org/10.1016/j.ijinfomgt.2015.07.008

Dutton, H.W.: Putting things to work: social and policy challenges for the internet of things. Info **16**(3), 1–21 (2014)

Funaki, K.: Customer Co-creation to Deliver Innovation in the Digital Era. Collaborative Creation Through Global R&D **66**(6) (2017)

Gebauer, H., Joncourt, S., Saul, C.: Services in product-oriented companies:past, present, and future. Universia Bus. Rev. **49**, 32–53 (2016)

Glova, J., Sabol, T., Vajda, V.: Business models for the internet of things environment. Procedia Econ. Finan. **15**(14), 1122–1129 (2014). https://doi.org/10.1016/S2212-5671(14)00566-8

Herrera, M.E.B.: Creating competitive advantage by institutionalizing corporate social innovation. J. Bus. Res. **68**(7), 1468–1474 (2015). https://doi.org/10.1016/j.jbusres.2015.01.036

Joyce, A., Paquin, R.L.: The triple layered business model canvas: a tool to design more sustainable business models. J. Clean. Prod. **135**, 1474–1486 (2016). https://doi.org/10.1016/j.jclepro.2016.06.067

Kaldaru, H., Parts, E.: Social and institutional factors of economic development: evidence from Europe. Baltic J. Econ. **8**(1), 29–51 (2008). https://doi.org/10.1080/1406099X.2008.10840444

Koohborfardhaghighi, S., Altmann, J.: A network formation model for social object networks. In: Zhang, Z., Shen, Z.M., Zhang, J., Zhang, R. (eds.) LISS 2014, pp. 615–625. Springer, Heidelberg (2015). https://doi.org/10.1007/978-3-662-43871-8_89

Koohborfardhaghighi, S., Altmann, J.: How strategic networking impacts the networking outcome: a complex adaptive system approach. In: Proceedings of the 18th Annual International Conference on Electronic Commerce: e-Commerce in Smart Connected World, p. 30. ACM (2016a)

Koohborfardhaghighi, S., Altmann, J.: How network visibility and strategic networking leads to the emergence of certain network characteristics: a complex adaptive system approach. In: Proceedings of the 18th Annual International Conference on Electronic Commerce: e-Commerce in Smart Connected World, p. 29. ACM (2016b)

Koohborfardhaghighi, S., Lee, D.B., Kim, J.: A study on the connectivity patterns of individuals within an informal communication network. In: Park, J., Yi, G., Jeong, Y.S., Shen, H. (eds.) Advances in Parallel and Distributed Computing and Ubiquitous Services, pp. 161–166. Springer, Singapore (2016). https://doi.org/10.1007/978-981-10-0068-3_20

Koohborfardhaghighi, S., Lee, D.B., Kim, J.: How different connectivity patterns of individuals within an organization can speed up organizational learning. Multimedia Tools Appl. **76**(17), 17923–17936 (2017)

Li, S., Xu, L.Da, Zhao, S.: The internet of things: a survey. Inf. Syst. Front. **17**(2), 243–259 (2015). https://doi.org/10.1007/s10796-014-9492-7

Metallo, C., Agrifoglio, R., Schiavone, F., Mueller, J.: Understanding business model in the Internet of Things industry. Technological Forecasting and Social Change, February 2017, pp. 0–1 (2018). https://doi.org/10.1016/j.techfore.2018.01.020

Mulgan, G., Tucker, S., Ali, R., Sanders, B.: Social Innovation: What It Is, Why It Matters and How It Can Be Accelerated. Basingstoke Press, London (2007)

Rouse, M.: What is business model? - Definition from WhatIs.com (2013). http://whatis.techtarget.com/definition/business-model. Accessed Feb 2018

Shin, D.: A socio-technical framework for Internet-of-Things design: a human-centered design for the Internet of Things. Telematics Inform. **31**(4), 519–531 (2014)

Westerlund, M., Leminen, S., Rajahonka, M.: Designing Business Models for the Internet of Things. Technol. Innov. Manage. Rev. **4**(7), 5–14 (2014). https://doi.org/10.1007/978-3-642-19157-2

Zwetsloot, G.I.J.M., Scheppingen, A.R.V., Bos, E.H., Dijkman, A., Starren, A.: The core values that support health, safety, and well-being at work. Safety and Health at Work **4**(4), 187–196 (2013). https://doi.org/10.1016/j.shaw.2013.10.001

Special Topic Session - Machine Learning, Cognitive Systems and Data Science for System Management

FaaStest - Machine Learning Based Cost and Performance FaaS Optimization

Shay Horovitz[✉], Roei Amos, Ohad Baruch, Tomer Cohen,
Tal Oyar, and Afik Deri

School of Computer Science, College of Management Academic Studies,
Rishon LeZion, Israel
horovitz@colman.ac.il

Abstract. With the emergence of Function-as-a-Service (FaaS) in the
cloud, pay-per-use pricing models became available along with the traditional fixed price model for VMs and increased the complexity of selecting the optimal platform for a given service. We present FaaStest - an
autonomous solution for cost and performance optimization of FaaS services by taking a hybrid approach - learning the behavioral patterns of
the service and dynamically selecting the optimal platform. Moreover,
we combine a prediction based solution for reducing cold starts of FaaS
services. Experiments present a reduction of over 50% in cost and over
90% in response time for FaaS calls.

Keywords: Function as a Service · Serverless · Machine Learning

1 Introduction

Over the past few years the cloud landscape has grown significantly [11,22],
allowing enterprises to move their workloads to it [24,26]. Cloud providers offer
a wide range of compute services a customer can select from such as IaaS, CaaS,
& PaaS. Serverless - Function as a Service (FaaS) technology is an newer alternative offering [34]. The ability to run code on demand without infrastructure
provisioning, was first presented in 2006 with *Zimki*, followed by Amazon AWS
Lambda in 2014. FaaS technology continues to gain popularity and still is a
major hype expected to grow further [10,32].

Due to the nature of Serverless services and the machinery behind it, FaaS
services come with limitations. FaaS is limited for stateless services; once the
request is finished, the VM/Container/Process in the backend is dead along
with its memory. Coding languages to set the functionality is limited as well. In
addition, resources per function call are limited - memory is bounded to 3 GB,
code and dependencies must not exceed 250 MB. As such, Serverless technology usually does not hold by itself, but serves as a complementary for other
computing infrastructures as IAAS, CAAS and PAAS. Although Serverless may
sound a very appealing service offering specifically due to its pay-per-use payment method, in fact it can cause the total cost to become even higher than

© Springer Nature Switzerland AG 2019
M. Coppola et al. (Eds.): GECON 2018, LNCS 11113, pp. 171–186, 2019.
https://doi.org/10.1007/978-3-030-13342-9_15

legacy compute services, as in the case of functions that consume high memory, computationally heavy or frequently triggered.

Notwithstanding cost considerations there are additional challenges [23] when using FaaS, yet, two issues are most prominent; First, the complexity of handling multiple pricing models: Currently FaaS is mostly economical when CPU-bound computations are dominant, while I/O bound functions may be cheaper on VMs/containers, and for services with dynamically changing resource consumption it is impossible to tell in advance which compute technology is optimal. Further, in the typical case where function calls are not adjacent in time, comes a major decrease in performance due to latency caused by instantiating resources for the function after long time of inactivity, also known as FaaS "cold start" [28,29,31,33,36]. Hence, there is a need for a solution that will optimize the cost of a given application or service by leveraging the most cost effective compute platform at a certain time, dynamically. Moreover, it would be beneficial if FaaS cold starts could be reduced when possible.

In this paper we present FaaStest - a solution for Cost and Performance effective Function as a Service optimization using Machine Learning, by predicting upcoming function calls and optimizing the compute platform for each behavioral pattern. We've experimented on a real application with various behaviors, changing loads dynamically over time. Results shows a decrease of over 50% in cost. Likewise, FaaStest managed to predict future events of function calls with an average accuracy of 98% and achieved an improvement of over 100% in performance.

2   Related Work

The existence of multiple pricing models in cloud services promotes confusion [20]. Additional research was done on pricing aspects of serverless and various optimization solutions to tackle it. In [5,9] the authors provide comparisons of the Serverless service prices among various cloud providers (Google, Azure and AWS) and a comparison between the cost of Serverless versus VM (Virtual Machines). [35] presents a Cost comparison study of monolithic, customer based microservice and cloud provider based microservice architectures. [27,35] examines the dependency of serverless cost on the execution behavior and volumes of the application workload and specifically for the case of microservices. [25] provides tools to plan cloud hosting budget for reserved, on demand or serverless models. [18,21,23] provides analysis about the various factors that affect the cost in serverless based architectures.

As serverless gains popularity, progressively more publications [19,23] about the flaws of its performance are presented, due to the cold start upon infrequent calls to the function [36]. [3,4] suggests methods that can reduce the cold start effect, such as using dynamically typed languages instead of statically typed languages, avoid putting lambdas in VPC and consider allocating more memory for the function. [14,16,17] suggest to reduce the cold start by invoking the serverless function with a configured time interval and by it forcing the container to stay alive. The problem with this approach is that it produces many redundant function calls, resulting in substantial increase in expenses.

3   Problem

Serverless comes with 3 main issues: (1) Pay-per-use: Although it is probably the biggest reason why a customer would choose to be hosted on serverless [2], it may act exactly the opposite and cause the cost to increase. (2) Cold start: Invoking requests not too often will cause the application to start with all the dependencies around it [3] - ending with poor performance. (3) Vendor dependency: Applications are completely dependent on the Cloud Provider they reside on, from managing the infrastructure, debugging and monitoring the application, to their specific APIs. In this paper we focus on the cost problems derived from the pay-per-use pricing model and the performance of the function due to cold starts. We examined the above mentioned problems on Node Cellar [15] application.

3.1   Cost Efficiency

In Fig. 1(a), we ran 10M requests over time - representing a highly loaded application, divided evenly across the whole time period to see how that affects the cost of serverless. The figure presents an accumulated cost for Lambda (FaaS) and VM. Up to 2:45 h, the serverless solution was cheaper, but afterwards it

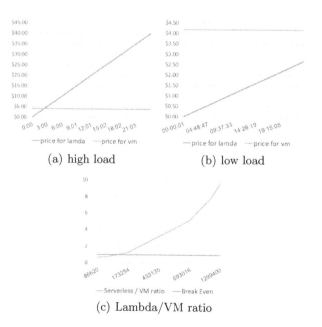

(a) high load   (b) low load

(c) Lambda/VM ratio

Fig. 1. Lambda vs VM pricing over time

became cost ineffective compared to the VM fixed pricing model. As depicted, such load may cause serverless prices to surpass a plain virtual machine by a factor of almost 10, which proves the importance of awareness in selecting the optimal compute platform for a given load. In contrast, in Fig. 1(b) we ran 1M requests over time - representing a low load; In this case it is clear that Lambda is the cost effective platform under those conditions.

In Fig. 1(c) we can see the Serverless/VM price ratio (Y axis) based on the amount of code execution time in 100-ms (X axis). Here we neglect the AWS Lambda free tier usage. This visualizes the break even point of our test where Serverless becomes more expensive than VM. However, per each application this break even point is different as it depends on the behavior of function calls.

As mentioned earlier, the Serverless formula is based on 3 main parameters - memory, execution time & number of requests. In Fig. 2 we examine the effect of each of those parameters in separate on the total cost over time. As the application gains more users and increased load, the VMs need to scale out in order to support that load and with it, the prices are going up.

While constant High usage vs a constant Low usage will have a great difference from a cost perspective and easy to fit the most cost effective compute platform, in real applications the loads tend to behave more dynamically, as

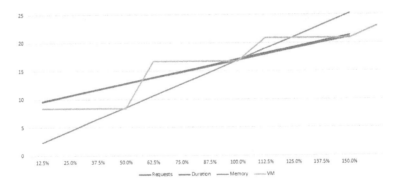

Fig. 2. Attributes effect on VM price

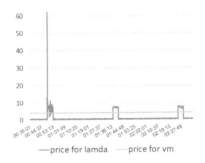

Fig. 3. FaaS vs VM on dynamic load

such we tested both solutions of VM and serverless for a real user-like behavior application. In this experiment using our JMeter application we ran 1.2M requests against our Nodecellar application, which were divided as such - each day we had a 2-h flash crowd of users, resulting in 20K requests total for those 2 h, and for the rest of the 22 h of the day - the rest of the 20K requests summing up to 40K requests a day and 1.2M requests a month accordingly. In Fig. 3 we can see a fragment of 3 days, whereas for the parts of the 2-h storms, the price for Lambda was higher than the price for a VM, and for the rest of the parts, Lambda price was lower than the VM.

3.2 Cold Start Impact on Performance

As for the cold start phenomenon [8], it derives from the fact that Serverless providers typically don't keep copies of your function running. Therefore, as pictured in Fig. 4 when a call has been made to a function that hasn't been active for a long time, it will require the function code to be started, and initializing different resources for hosting it [13]. This process will cause the function's response time to be dozens of times higher, dependent [12] on the serverless services provider, each one offering its own unique implementation, features and benefits.

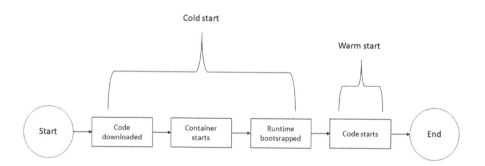

Fig. 4. FaaS cold start structure

We use Fission [6] - a Kubernetes cluster that function as Serverless and maintains a pool of "warm" containers. When a function is first called, i.e. "cold-started", a running container is chosen and the function is loaded. First, we looked at the cold start behavior within Fission and its effect on performance. We deployed a simple javascript function which returns a message containing a random number in order to avoid any caching. We call this function with different time gaps and measure its response time. As appear in the result in Fig. 5, in Fission it takes around 6 min for a function to get cold.

Then, we tested the effect of cpu load on the serverless function response time. We used CPUSTRESS tool which enabled us to keep the cpu busy steadily on 95% usage. In parallel, using JMeter we executed calls, with time interval of

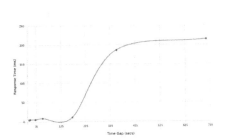

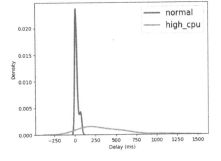

Fig. 5. FaaS response time for different time gaps

Fig. 6. CPU load effect on cold start time

3 min and recorded the response time. Figure 6 shows the distribution of response time in both normal and stressed CPU. In normal mode we experienced only a few milliseconds of response time, while in the high cpu mode it risen up mostly to 250 ms, and in many occasions it was even much worse.

4   Solution

In high level, FaaStest monitors the function call events of a given application, and dynamically selects the optimal compute platform for the current call behavioral pattern such that the total cost is reduced. It collects events of function calls from the past in a time series, and analyzes the cost of each window per each platform - FaaS or VM. Then it runs a classification algorithm in order to identify at a given time whether FaaS of VM would be cost efficient. In parallel, for the time ranges that FaaStest decides to use FaaS platform, it will run a prediction algorithm for the estimated time of a function call, and warm up the service prior to the actual call of the function, preventing its cold start effect on performance.

4.1   Platform Automatic Selection Based Cost Reduction

The architecture of our solution depicted in Fig. 7 is based on a sliding window oriented approach, which includes several key elements required to achieve the most optimal price for a specific application, these key elements are:

JMeter - allows us to imitate user's requests with dynamically frequent function call patterns. **Node Cellar** - A web application. **New Relic Service** - Used by the New Relic (APM solution for collecting performance metrics) agent installed on the Node Cellar virtual machine. **New Relic Export** - Exporting the gathered data by New Relic to a local store, in a sliding window, for prediction by the model in *Data To Predict*, and eventually to create the training data in *Training Data*. **Pre Processing** - Transforming the Request Data into Training Data by cleaning the data, compressing and formatting it per minute.

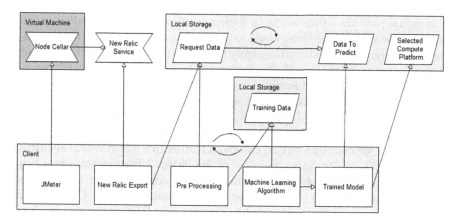

Fig. 7. Automatic platform selection in FaaStest

Machine Learning Algorithm - Decision Tree classification based algorithm, training our model based on two thirds of the Training Data and testing it on one third of the data. **Trained Model** - Using our trained model from the Machine Learning Algorithm phase, and based on the data To predict from the New Relic Export phase, create a new dataset which includes the most optimal compute service in the Selected Compute Platform section.

4.2   Cost Function

In order to assess the expected cost per each behavior of the application function calls at a given time, FaaStest computes the current cost based on the behavior of the latest sliding window. We refer to the form of payment for a single non-scalable VM with 512 MB of RAM, which its monthly cost is \$4.18, as published in [1], May 2018. The VM calculation function is fairly simple - the monthly cost of the VM is divided by the number of days in the month, hours, and seconds as the following: \$4.18/(30∗24∗60∗60). As for FaaS, we used the formula published in [5] as the following:

$$P_T = (NT_E*(M/1024)-Q_F)*P_G+(NT_E*(C/1000)-P_F)*P_C+(N-N_F)*P_R$$

where: N is number of executions in month. N_F is number of free requests per month. P_G is price per GB-second. P_C is price per GHz-second. P_r is price per request. P_T is TOTAL monthly price. T_E is time per execution in seconds. M is memory allocation in MB. C is CPU allocation in MHz. Q_F is free GB-seconds and P_F is free GHz-seconds.

4.3   Features for Learning Multi-frequency Function Call Seasonality

In order to trace for seasonality of the function calls, we examined time series algorithms such as ARIMA which failed to model the behavior of our function call seasonal patterns - that is due to having multiple mixed seasonal patterns of function calls at the same time, along with function calls that are not seasonal in behavior. Running a classification algorithm with time features such as "Hour", "Minute" and "Second" would also fail since the period between calls may not correlate with 60 min. Yet, having multiple mixed seasonal patterns is even more natural and likely to happen in real applications. As such, we took a hybrid approach: first we ran DFT (Discrete Fourier Transform) on function call events, in order to identify the bold frequencies of function inter-call time. Then - for each frequency we created a virtual clock as a feature in a classification algorithm.

For example, in Fig. 8, in case (a) we emulated 2 frequencies - of 19 and 37 min between function calls and embedded 1% noise - calling the functions at random times; in case (b) we emulated 3 frequences - of 29, 41 and 49 min between function calls and embedded 10% noise. The DFT found the dominant frequencies successfully, yet as there's more noise and more frequencies - we can see that additional spikes appear in the figure - either due to harmonies and interference of the dominant frequencies and due to the noise. In order to eliminate those spikes, we assign for each found frequency a virtual clock - a counter that counts up to modulu the frequency size; All clocks are set as features for the input of our classification algorithm, such that it can identify seasonal patterns of multi-frequency function calls. As the classification algorithm selects the optimal set of features that allows it to create a model for the data, it automatically neglects the frequencies that are caused by interference or noise.

It is also possible, even though not implemented in our experiment, to extract the parameters of each function call as features as well such that the classification algorithm can potentially identify relations between the frequency and a subset of parameter values of the function.

We focused our tests on backend processes, yet for frontend processes (user processes), when the major time-dependent feature is the load on the server, ARIMA would work fine, or even a simpler time representation as features of: minute, hour, day, month & year. For the classification algorithm we tested

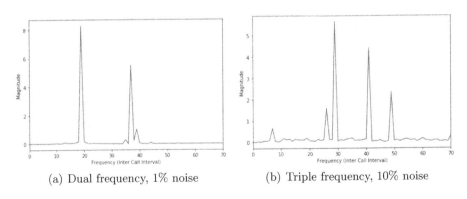

(a) Dual frequency, 1% noise   (b) Triple frequency, 10% noise

Fig. 8. DFT for FaaS function call time frequencies

several alternatives, and test results appear in Sect. 5. Finally we used Decision Tree due to its high accuracy and our ability to examine the model in depth.

4.4   Cold Start Reduction

Once we have automatically selected the optimal compute platform, in the case that our algorithm decided to use the Serverless platform, the application may face performance degradation due to the cold start problem, and following we describe our solution - based on predicting the time of the next function call using historical function calls data. This allows to invoke the function just before each of the predicted event times.

Testing Environment: Our experiments were conducted using the Fission framework for serverless functions [6]. Fission uses Kubernetes as an infrastructure for the function-as-a-service architecture. In order to run kubernetes, we used MiniKube, which is a tool for running Kubernetes. MiniKube runs a single-node Kubernetes cluster inside a virtual machine.

For the sake of the experiments conducted, we had to alter the source code of Fission so that we can reduce the gap that cold start appears in. After these adjustments has been made, the gap required for cold start to appear was reduced from 6 to 1 min, this change was made in order to enable us performing more experiments in a shorter period of time. All experiments were performed on a Lenovo yoga 710 with 8 GB ram, 256 GB ssd disk, with an Intel i7 - 7500U cpu.

Testing Application: As depicted in Fig. 9, we use Nodecellar in all of our experiments - a basic web application based on the Express framework with a Node.js server side. As shown in the diagram above, the process of collecting the data needed for training our prediction model of upcoming serverless function calls, starts with simulating user activity in our node cellar application. Apache JMeter was used for simulating user's actions in our client application, the simulation was done by performing http requests directly to the server side of the application. We simulated these user actions in a configured interval of 1 min. We altered the server side code of the node cellar application so that for each

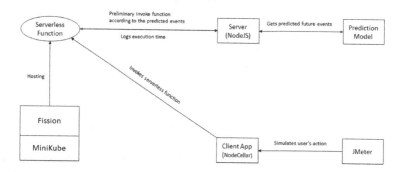

Fig. 9. FaaS cold start reduction process elements

of the performed user actions, it performed another http request to our hosted serverless function and invoked it. A separate Node.js server, was responsible to log the activity of this serverless function. Each time it received a request from the serverless function - it logged the executed function's name and execution time to a local csv file. Now proceed to the code of our serverless function.

The code we have chosen to host on the fission platform, was consisted of two parts. The first one was performing a new http request to the previously mentioned Node.js server in order to log the function execution time. The second part was generating a random number and returning it to client application, this was done in order to avoid any caching situations. After simulating user actions that led to invoking our serverless function and logging the execution time of function, we were able to create our dataset. The dataset we used for training our prediction model was based on the csv file used by our Node.js to log the function activity. The dataset was set having two columns: $requestTime$ column-is the time when the function was invoked, with resolution of seconds and call column contains value that can be either 1 or 0, 1 indicates that function call has been made in that specific second, otherwise the value will be zero. Our goal was to try and detect the seasonality in our calls, i.e. interval of 1 min. Our Node.js was also in charge of executing python script for predicting upcoming function calls. Our python code read that data, splitted the $requestTime$ column to 3 different columns - hour, minute, second. Than we added 1% noise to our dataset, which was reflected with 1% of the rows had their value column set to "1". The next step was trying to detect the seasonality in the noisy dataset and predict the time when future events will occur. The accuracy, precision, recall and AUC of the tests we conducted appear in Sect. 5. After the seasonality mode has been learnt, we use it in order to predict when to call the function prior to its anticipated calling time - and this way prevent its cold start. While the extra call to the function also increase the total cost, in a real environment where the service provider utilizes our solution, it can simply warm up the function without the penalty of this cost - which is part of the pricing model. After receiving all of the prediction made by our model, the Node.js server had set different timers for each of the predicted upcoming function calls. When each of these timers came to its end, it performed a new http call to our serverless function. By doing so, when the actual call was made by the user just a few seconds later - the serverless function was already up and running and was able to respond immediately. The timers was set 10 s before the predicted call time - this was done to guarantee that fission platform has sufficient time to complete the function's initialization and running phase before the predicted call was actually made.

5   Experiments

In order to achieve the above solution, we conducted the following experiments. For our experiments we deployed a Node Cellar application in the cloud with a New Relic agent installed on it, and we used JMeter to mimic real users behavior. The same infrastructure described in Sect. 4 chapter was used for the following experiment.

5.1   Classification Algorithms Accuracy

The classification machine algorithms we employed were taken from the Python Scikit Learn Machine Learning algorithms library.

In the testing of FaaStest we worked in four stages, the first was to create our raw data to manipulate, second was to find the correct label for each sample, third was to find the best features that would give us the highest percentage of accuracy using our mixed DFT and classification solution, and fourth was to test each algorithm to see which one has the highest accuracy.

To create our testing set we ran 1.2 Million requests in total within 24 h - emulating traffic for 30 days, i.e. the features were collected per minute (which represents 30 min in real life) - ending with 1440 samples for the test. Per each minute of data, the feature vector used for learning contains the following: hour of day, minute within that hour, virtual clock features (As explained in Sect. 4.3) and a label with the value of 0 for serverless and 1 for VM. Per each minute, we use the cost function formula (in Sect. 4.2) in order to compute the price of serverless for that minute (30 min in real life) of requests. Then this price is compared with the fixed price of VM. The platform (either VM of serverless) with the minimal price is written as the label for that specific minute feature vector. The minute-based vectors were divided into training set and validation and testing sets with 2:1:1 ratio between the training and the validation and testing sets. Different parameters were used for validating the best model with highest AUC and Accuracy rates.

In Fig. 10 we tested the Accuracy, Precision, Recall and AUC of the classification in the case of prediction of the least cost platform over time. For our pricing optimization algorithm, Decision Tree received the highest Accuracy score. The most influencing parameters in our calculations were: max_depth - depth of the decision tree, criterion - gini impurity or Entropy and $class\_weight$ - to indicate if some classes have higher weights than others. In Fig. 11 we tested the Accuracy, Precision, Recall and AUC of the classification in the case of Cold Start prevention, predicting the occurrence of FaaS function calls only. SVC with RBF kernel and Decision Tree achieved the best results overall.

Model	Decision Tree	SVC	Naïve Bayes	MLP	KNN
Parameters used	max_depth=None, criterion="Gini", class weight=None	C=0.025, kernel=Poly, degree=1	alpha=1.0, fit_prior=True	alpha=0.0001, activation="relu", solver="adam"	n_neighbours=3, weights=uniform, p=2(euclidean distance)
AUC	86.22%	65.21%	57.16%	50.09%	71.08%
Precision	76.66%	12.28%	11.45%	16.66%	53.84%
Recall	74.19%	67.74%	35.48%	3.22%	45.16%
Accuracy	96.53%	63.04%	75.75%	91.91%	93.30%

Model	Decision Tree	SVC	Naïve Bayes	MLP	KNN
Parameters used	max_depth=16, criterion="Gini", class weight=None	C=0.025, kernel=Pol y, degree=3	Changing the alpha didn't affect the score	alpha=0.00 01, activation= "tanh", solver="ad am"	n_neighbours=3, weights=distance p=2(euclidean distance)
AUC	86.35%	71.33%		54.71%	70.96%
Precision	79.31%	39.47%		75.00%	51.85%
Recall	74.19%	48.38%		9.67%	45.16%
Accuracy	97.76%	90.99%		93.30%	93.07%

Model	Decision Tree	SVC	Naïve Bayes	MLP	KNN
Parameters used	max_depth=8, criterion="Gini", class weight=None	C=0.025, kernel=Poly, degree=2	alpha=1.0, fit_prior=False	alpha=0.01, activation="relu", solver="adam"	n_neighbours=5, weights=uniform, p=2(euclidean distance)
AUC	78.03%	66.99%	66.95%	54.83%	58.55%
Precision	69.23%	38.70%	13.37%	23.52%	40.00%
Recall	58.06%	38.70%	67.74%	12.90%	19.35%
Accuracy	95.15%	91.22%	66.28%	90.76%	92.14%

Model	Decision Tree	SVC	Naïve Bayes	MLP	KNN
Parameters used	max_depth=None criterion="Entrop y", class weight=None	C=0.025, kernel=Pol y, degree=4	Changing the alpha didn't affect the score	alpha=0.00 01, activation= "relu", solver="sg d"	n_neighbours=3, weights=distance p=2(euclidean distance)
AUC	89.45%	70.95%		48.87%	66.87%
Precision	78.12%	36.58%		4.34%	61.11%
Recall	80.64%	48.38%		3.22%	35.48%
Accuracy	96.99%	90.30%		87.99%	93.76%

Fig. 10. Platform selection prediction algorithm performance

Model	DecisionTree	NaiveBayes	SVC	SGD	RBF_SVC
Parameters used	max_depth=None, Criterion='Gini', class weight=None	alpha=1.0, fit_prior=True	C=0.025, kernel=poly, degree=1	penalty=none, loss='hinge'	C=1, kernel='rbf', degree =1
Accuracy	98.90%	80.98%	97.63%	97.52%	99.40%
Precision	74.44%	9.93%	0.00%	0.00%	100.00%
Recall	77.68%	84.94%	0.00%	0.00%	76.76%
AUC	88.53%	82.91%	50.00%	50.00%	88.38%

Model	DecisionTree	NaiveBayes	SVC	SGD	RBF_SVC
Parameters used	max_depth=16, Criterion='Gini', class weight=None	changing the alpha didn't affect the score	C=0.025, kernel=poly, degree=2	penalty=l1, loss='hinge'	C=1, kernel='rbf', degree =2
Accuracy	99.01%		97.50%	96.32%	99.47%
Precision	78.57%		0.00%	38.30%	100.00%
Recall	81.91%		0.00%	81.91%	79.16%
AUC	90.67%		50.00%	89.30%	89.58%

Model	DecisionTree	NaiveBayes	SVC	SGD	RBF_SVC
Parameters used	max_depth=8, Criterion='Gini', class weight=None	changing the alpha didn't affect the score	C=0.025, kernel=poly, degree=3	penalty=l2, loss='hinge'	C=1, kernel='rbf', degree=3
Accuracy	99.45%		98.90%	96.09%	99.45%
Precision	98.70%		100.00%	37.37%	100.00%
Recall	79.16%		58.58%	78.57%	78.35%
AUC	89.56%		79.29%	87.56%	89.17%

Model	DecisionTree	NaiveBayes	SVC	SGD	RBF_SVC
Parameters used	max_depth=8, Criterion='Gini', class weight=None	alpha=1.0, fit_prior=False	C=0.025, kernel=poly, degree=4	penalty=elasticnet, loss='hinge'	C=1, kernel='rbf', degree =4
Accuracy	98.75%	77.18%	99.63%	97.60%	99.55%
Precision	72.89%	8.59%	100.00%	100.00%	100.00%
Recall	80.41%	84.37%	84.44%	5.15%	81.72%
AUC	89.81%	80.68%	92.22%	52.57%	90.86%

Fig. 11. FaaS function call time prediction algorithm performance

5.2   Platform Automatic Selection

In Fig. 12 we compared our machine learning based prediction with 3 alternatives: Lambda (FaaS only solution), VM, Ideal solution and real time prediction. Lambda Only represents the cost model of a FaaS only based solution, according to the pricing of Amazon Lambda. VM Only represents the cost model of a VM only solution, according to the pricing of Amazon EC2. The Real Time simple prediction algorithm calculates the most cost effective method to use for the next hour based on statistical analysis of the last 4 h. The Ideal presented in this figure represents the optimal in theory where one would select the best platform every second, and the Machine Learning is our FaaStest proposed solution. The graph shows the cumulative price of each of the above over time. The load of the application was dynamic with the same settings we used in Sect. 3. FaasTest presents better results than the alternatives, in terms of pricing.

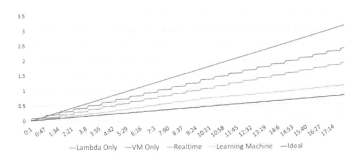

Fig. 12. FaaStest vs alternatives cumulative price over time

5.3   Cold Start Prevention

In Fig. 13(a) we experiment the compare our FaaS function call prediction with the common solution proposed by [14,16,17]. We aim to compare the cost of our solution - predicting future events and make preliminary calls, to the existing

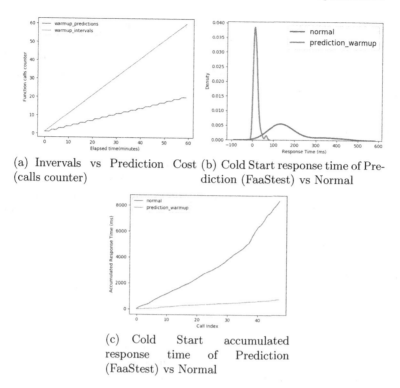

(a) Invervals vs Prediction Cost
(calls counter)

(b) Cold Start response time of Prediction (FaaStest) vs Normal

(c) Cold Start accumulated response time of Prediction (FaaStest) vs Normal

Fig. 13. Cold start prevention algorithm performance

common solution - set fixed intervals that will warm up the function. We set an interval of 1 min. Using JMeter we simulated calls 3 min away from each other, we simulated 20 calls, so in total the simulation took 60 min. We measured the amount of preliminary calls each of the methods made to warm up the function before the upcoming simulated call. The result, was a drastic decrease in the amount of calls that had to be made using our solution, leading to saving of over 50% in the total cost.

In Fig. 13(b) we compared the response time received from the tested serverless function in the normal situation, where no interferences were made, against the response time achieved when using our solution - future events predictions followed by preliminary warm up calls. This graph shows the response time distribution in both of the tested situations. We can see the cold start phenomenon occurring in the normal situation, with a response time with average of 175 ms and stddev of 95. In the case of our FaaStest, the preliminary warm up calls made merely prevented cold start from occurring, with response time average of 16.4 ms and stddev of 11.8.

In Fig. 13(c), we visualized the same dataset of the last experiment, response time achieved in the normal situation against response of our solution, but in a cumulative manner. We can see a major difference between the accumulated waiting time a real user would experience in the current situation against the time he will wait in a system using FaaStest.

6   Summary

In this paper we presented FaaStest - a cost and performance effective Function as a Service optimization using Machine Learning. FaaStest brings a novel hybrid approach for selecting the optimal Function service platform dynamically in accordance to ML-learned function calls behavioral patterns. In experiments this approach gained over 50% reduction in cost. In addition, once Serverless has been chosen through the proper model, we were able to overcome the Cold Start problem using a prediction based solution with a total of 90% reduction in average response time.

References

1. Amazon On-Demand Pricing. https://aws.amazon.com/ec2/pricing/
2. Amazon Web Services pricing. https://aws.amazon.com/lambda/pricing/
3. Dealing with cold starts in AWS Lambda. https://medium.com/thundra/dealing-with-cold-starts-in-aws-lambda-a5e3aa8f532
4. Does coding language memory or package size affects colds tarts of AWS Lambda. https://read.acloud.guru/does-coding-language-memory-or-package-size-affect-cold-starts-of-aws-lambda-a15e26d12c76
5. Economics of serverless computing. https://451research.com/report-long?icid=4406?utm_source=trending_topics&utm_term=cloud_pricing
6. Fission official documentation. https://docs.fission.io/0.7.2/
7. Fission serverless function as a service for kubernetes. https://kubernetes.io/blog/2017/01/fission-serverless-functions-as-service-for-kubernetes/
8. From Containers to AWS Lambda. https://blog.travelex.io/from-containers-to-aws-lambda-23f712f9e925
9. Function as a Service (FaaS) - why you should care and what you need to know. https://www.redhat.com/files/summit/session-assets/2017/S109151-serverless.pdf
10. Function-as-a-Service Market Global Forecast to 2021. https://www.researchandmarkets.com/research/nfq5pr/functionasaserv
11. Gartner Forecasts Worldwide Public Cloud Services Revenue to Reach $260 Billion in 2017. https://www.gartner.com/newsroom/id/3815165
12. Get functional! 5 open source frameworks for serverless computing. https://www.infoworld.com/article/3193119/open-source-tools/get-functional-5-open-source-frameworks-for-serverless-computing.html
13. Go Serverless - pros and cons. https://devops.com/go-serverless-pros-cons/
14. Keep your lambdas warm (interval based solution). https://serverless.com/blog/keep-your-lambdas-warm/
15. Node Cellar source code. https://github.com/ccoenraets/nodecellar

16. Open source project suggesting Warmup support for AWS Lambda functions to prevent cold starts. https://github.com/thundra-io/thundra-lambda-warmup/blob/master/README.md

17. Resolving cold start in AWS Lambda. https://medium.com/@lakshmanLD/resolving-cold-start

18. Serverless, a new cloud trend. https://medium.com/slalom-engineering/serverless-the-new-cloud-trend-e2f163433431

19. Serverless challenges (cold start). https://hackernoon.com/the-key-challenges-serverless-will-have-to-overcome-to-succeed-in-2018-af3132ed4995

20. The Financial Case for Moving to the Cloud. gartner.com/smarterwithgartner/the-financial-case-for-moving-to-the-cloud/

21. The hidden costs of Serverless. https://medium.com/@amiram_26122/the-hidden-costs-of-serverless-6ced7844780b

22. Worldwide Public Cloud Services Spending Forecast to Reach $160 Billion This Year, According to IDC. https://www.idc.com/getdoc.jsp?containerId=prUS43511618

23. Baldini, I., et al.: Serverless computing: current trends and open problems. In: Chaudhary, S., Somani, G., Buyya, R. (eds.) Research Advances in Cloud Computing, pp. 1–20. Springer, Singapore (2017). https://doi.org/10.1007/978-981-10-5026-8_1

24. Bhattacherjee, A., Park, S.C.: Why end-users move to the cloud: a migration-theoretic analysis. Eur. J. Inf. Syst. 23(3), 357–372 (2014)

25. Boza, E.F., Abad, C.L., Villavicencio, M., Quimba, S., Plaza, J.A.: Reserved, on demand or serverless: model-based simulations for cloud budget planning. In: 2017 IEEE Ecuador Technical Chapters Meeting (ETCM), pp. 1–6. IEEE (2017)

26. Buyya, R., Yeo, C.S., Venugopal, S., Broberg, J., Brandic, I.: Cloud computing and emerging it platforms: vision, hype, and reality for delivering computing as the 5th utility. Future Gener. Comput. Syst. 25(6), 599–616 (2009)

27. Eivy, A.: Be wary of the economics of "serverless" cloud computing. IEEE Cloud Comput. 4(2), 6–12 (2017)

28. Hendrickson, S., Sturdevant, S., Harter, T., Venkataramani, V., Arpaci-Dusseau, A.C., Arpaci-Dusseau, R.H.: Serverless computation with OpenLambda. Elastic 60, 80 (2016)

29. Lee, H., Satyam, K., Fox, G.C.: Evaluation of production serverless computing environments. In: 3rd International Workshop on Serverless Computing (WoSC) (2018)

30. Lynn, T., Rosati, P., Lejeune, A., Emeakaroha, V.: A preliminary review of enterprise serverless cloud computing (Function-as-a-Service) platforms. In: 2017 IEEE International Conference on Cloud Computing Technology and Science (CloudCom), pp. 162–169. IEEE (2017)

31. Oakes, E., Yang, L., Houck, K., Harter, T., Arpaci-Dusseau, A.C., Arpaci-Dusseau, R.H.: Pipsqueak: lean Lambdas with large libraries. In: 2017 IEEE 37th International Conference on Distributed Computing Systems Workshops (ICDCSW), pp. 395–400. IEEE (2017)

32. Panetta, K.: Top trends in the gartner hype cycle for emerging technologies (2017). https://www.gartner.com/smarterwithgartner/top-trends-in-the-gartner-hype-cycle-for-emerging-technologies-2017/

33. Shillaker, S.: A provider-friendly serverless framework for latency-critical applications. In: 12th Eurosys Doctoral Workshop, Porto, Portugal (2018)

34. Varghese, B., Buyya, R.: Next generation cloud computing: new trends and research directions. Future Gener. Comput. Syst. 79, 849–861 (2018)

35. Villamizar, M., et al.: Cost comparison of running web applications in the cloud using monolithic, microservice, and AWS Lambda architectures. SOCA **11**(2), 233–247 (2017)
36. Wang, L., Li, M., Zhang, Y., Ristenpart, T., Swift, M.: Peeking behind the curtains of serverless platforms. In: 2018 USENIX Annual Technical Conference (USENIX ATC 2018), pp. 133–146. USENIX Association (2018)

Secure Query Processing over Encrypted Data Using a Distributed Index Structure for Outsourcing Sensitive Data

Hyunjo Lee, Hyeonguk Ma, Youngho Song, and Jae-Woo Chang$^{(\boxtimes)}$

Chonbuk National University, Jeonju, Republic of Korea
{o2near, akgusrnrl23, songyoungho, jwchang}@jbnu.ac.kr

Abstract. As the outsourcing of sensitive data has been spotlighted, data encryption schemes are required to protect the data. Accordingly, it is necessary to develop not only a distributed index structure to efficiently manage the large amount of encrypted data, but also a query processing scheme over the encrypted data. Meanwhile, the existing query processing schemes over the encrypted data cannot support top-k query processing algorithm which aim to quickly retrieve k number of the highest ranking tuples. To solve the problems, in this paper, we propose a secure query processing scheme over the encrypted data using a distributed index structure. The proposed distributed index structure guarantees data privacy preservation and performance improvement for the various types of queries. Finally, we show from our performance analysis that our proposed index structure and secure query processing scheme are suitable for protecting the data privacy of the sensitive data.

Keywords: Encrypted query processing · Distributed index structure · Data privacy

1 Introduction

Recently, banks and IT enterprises pay attention to analyzing a large amount of sensitive data which are received through internet banking and social networking services (SNS). To manage their own data, the companies deploy an infrastructure that requires database professionals who can set up physical hardware, install software, upgrade a system and keep the data up-to-date. This leads to a huge amount of expense on activities, deployment and operations related with data management. As a result, small-sized enterprises cannot establish a system to analyze large data because a huge amount of computing resources is required.

Therefore, researches on data outsourcing have been spotlighted. In data outsourcing, a data owner (DO) outsources his/her data to a service provider (SP) who is in charge of processing a query sent from an authorized user (AU). By outsourcing the data, the DO (e.g., the individuals or enterprises) can reduce the costs of initial investment and maintenance on computing infrastructure. According to the research on

© Springer Nature Switzerland AG 2019
M. Coppola et al. (Eds.): GECON 2018, LNCS 11113, pp. 187–198, 2019.
https://doi.org/10.1007/978-3-030-13342-9_16

outsourcing database market forecast in 2016[1], the revenue generated by service providers of outsourcing databases will be $1.8 billion, which is twelve times of the revenue generated in 2012 which is $150 million.

However, data outsourcing may cause the disclosure of some sensitive data by the SP, such as social security number and financial information. Therefore, the sensitive data should be protected to preserve the privacy of data. Meanwhile, the amount of sensitive data is rapidly increasing through internet banking and social networking services (i.e., Facebook and Twitter). For example, Facebook generates 4 new petabytes of data and runs 600,000 queries and 1 million map-reduce jobs per day [1].

For outsourcing sensitive data, an efficient data privacy preservation scheme is required. Accordingly, the DO can encrypt the original data before outsourcing sensitive data to the SP. The SP stores the data by constructing an index and returns a query result when the AU requests a query. However, to answer the query, the original data may be revealed to the SP and the query processing cost is high because the whole database should be decrypted by the SP [2, 3]. Therefore, it is necessary to process a query over the encrypted data to improve both data privacy and query processing performance.

To protect sensitive data in outsourced database environment, it is necessary to develop not only an efficient distributed index structure, but also a query processing scheme over the encrypted data. For this, Popa et al. proposed CryptDB [4]. CryptDB is a typical query processing scheme over the encrypted data. CryptDB encrypts data in a column-wise way by considering various query types, i.e., exact matching, range query and so on. To support the exact matching query, for example, it encrypts data using a deterministic encryption method such as AES. However, CryptDB cannot support top-k query processing algorithm. Because the top-k query is one of the useful one for data analysis, it is important to provide a top-k query processing algorithm over encrypted database.

To solve the problems, in this paper, we propose a distributed index structure and a secure query processing scheme for the encrypted data. Our contributions can be summarized as follows:

– We propose a distributed index structure for preserving the privacy of sensitive data. Our index structure consists of a prefix-based upper index structure to protect data partitioning information and a signature-based lower index structure to improve query processing performance.
– We propose a secure query processing scheme over the encrypted data, which can support not only arithmetic and comparison operations among data with different columns, but also a top-k search operation to find the k most relevant result.
– We also present an extensive experimental analysis of our query processing scheme by comparing it with the existing CryptDB for exact and range queries and the LPTA+ for top-k queries.
– We show that our query processing scheme provides both higher query processing performance and higher query result accuracy while preserving data privacy.

---

[1] https://451research.com/report-short?entityId=78105&referrer=marketing.

The rest of the paper is organized as follows. Section 2 presents the overall system architecture for data outsourcing. We explain the proposed distributed index structure and our secure query processing algorithm over the encrypted data in Sects. 3 and 4, respectively. In Sect. 5, we compare the performance of our query processing scheme with that of the existing schemes. In Sect. 6, we introduce the existing distributed index structure to deal with large data and the existing query processing schemes over the encrypted data. Finally, we conclude this paper with future work in Sect. 7.

2 System Architecture for Outsourcing Sensitive Data

Figure 1 shows the system architecture for outsourcing data on cloud computing. To preserve the privacy of the sensitive data, we propose not only a prefix-tree based distributed index structure for encrypted data, but also a secure query processing scheme over the encrypted data. The overall procedure for outsourcing sensitive data to the cloud computing is as follows. First, a data owner (DO) encrypts an original database and makes signatures to preserve data privacy. DO also generates keys for an upper index structure to distribute the encrypted data. Second, the DO sends the encrypted database, the signatures and the keys to a service provider (SP). Thirdly, the SP constructs a prefix-tree based upper index structure by using the keys sent from the DO. Fourthly, the SP stores the encrypted database and the signatures in a distributed manner. Fifthly, an authorized user (AU) encrypts a query and sends it to the SP. Sixthly, the SP processes the query by considering a query type and re-turns a candidate query result to the AU. Finally, the AU decrypts the candidate query result and obtains the final query result.

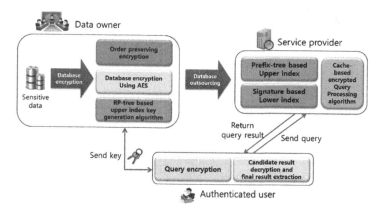

Fig. 1. System architecture for outsourcing sensitive data

3   Proposed Distributed Index Structure for Encrypted Data

3.1   Overall Architecture

Figure 2 shows the overall architecture of the proposed distributed index for the encrypted data. To enhance the performance for query processing over the large-capacity encrypted data, it is necessary to store data in a distributed manner. The proposed distributed index consists of three components: (i) bitmap key generation, (ii) a prefix-tree based upper index, (iii) a signature-based lower index. The algorithm for constructing our distributed index is as follows. First, a bitmap key of the encrypted data is generated by using Random Projection tree (RP-tree) [5]. RP-tree is a simple variant of k-d trees that adapts to intrinsic low dimension. In RP-tree, instead of splitting along coordinate directions at the median, it splits along a random direction in S^{k-1} with the unit sphere in R^k, where k means a number of dimensions. The algorithm generates a bitmap key by using RP-tree. We can protect data partitioning information from an attacker because RP-tree performs clustering by projecting data to an arbitrary partition axis. Then the algorithm encrypts the original data by using the AES scheme [2]. For query processing over the encrypted data, the signature of the original data is generated by using Order Preserving Encryption Scheme (OPES). Here, OPES is a deterministic encryption scheme whose encryption function preserves numerical ordering of the plaintexts. So, OPES allows operations, such as comparison, equality, range, MAX, MIN, COUNT, GROUP BY and ORDER BY, to be directly applied on encrypted data. As a result, DO sends to SP the encrypted record, i.e., <bitmap key, OPES encrypted signature, AES encrypted data>. Second, the SP extracts k number of front bits in the OPES encrypted signature and generates a hash table for the extracted front bits. Then, SP constructs a prefix-tree [6] and makes the connection between a hash table entry and a tree node. Finally, by using the upper index, the algorithm stores the data <OPES encrypted signature, AES encrypted data> into a node. In case when we maintain signatures with a cache, we can achieve the better query processing performance.

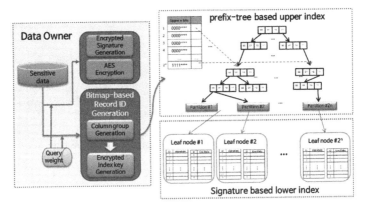

Fig. 2. Overall architecture of the proposed distributed index structure

3.2 Signature-Based Lower Index

A data node in the signature-based lower index contains a data <encrypted signature, AES encrypted data>. Because the size of the encrypted signature is much smaller than that of the AES encrypted data, we can store the encrypted signature into a cache, leading to the better retrieval performance by reducing the cost of disk I/O. Thus, we can determine the encrypted signatures to be cached by considering both their frequency and their recency. For this, we define a validity of recency (V_{Rec}) and a value of frequency (V_{Fre}). The V_{Rec} can be computed by measuring how much time is elapsed since a recent query has accessed the encrypted signatures. The V_{Fre} can be computed by the number of accesses to the encrypted signature in a cache for a specific period. The cache ratio using both V_{Rec} and V_{Fre} is calculated by using Eq. (1). Here, i means the issued/cashed query, t_{cur} is the current time and *timeInterval* means the elapsed time from the recent query time to t_{cur}. Count measures the number of queries issued from $t = 0$ to t_{cur}. And w_r and w_f mean the weights of recency and frequency, respectively.

$$V_{Rec}(i, t_{cur}) = V_{Rec}(i, t_{cur-1}) + t_{cur} \times \sqrt{\frac{t_{cur} - timeInterval}{2}}$$

$$V_{Fre}(i, t_{cur}) = \frac{Count(i, t_{cur}) - Count(i, t_{cur-1})}{timeInverval} \tag{1}$$

$$w_{cache}(i) = w_r \times V_{Rec}(i, t_{cur}) + w_f \times V_{Fre}(i, t_{cur})$$

4 Secure Query Processing Scheme over Encrypted Data

Figure 3 shows the overall architecture of the proposed secure query processing scheme over the encrypted data. First, when an authorized user (AU) requests a service, a data owner (DO) not only generates an upper index key using the RP-tree, but also

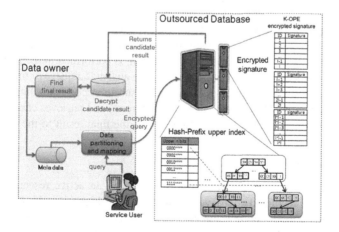

Fig. 3. Overall architecture of the proposed query processing scheme

creates the encrypted signature for a given query. Second, a service provider (SP) selects the data nodes corresponding to the upper index key for processing the query. Then, both the encrypted signature and the query are sent to the data nodes being selected. Third, the nodes retrieve the candidate results by performing the bit operations of the encrypted signatures. Fourth, the SP returns the encrypted candidate result to the DO. Fifth, the DO decrypts it and obtains the final result by filtering out unnecessary candidates. Finally, the DO sends the final result to the AU. Meanwhile, it is costly to retrieve Ng (# of column pairs) number of RP-trees when generating the upper index key. To enhance retrieval performance, we pre-construct a grid-tree mapping index for each partition in a column pair.

4.1   Exact-Match Query Processing

The exact-match query processing algorithm is performed as follows. First, the algorithm generates a b-bit key for a given query by using the grid-tree index and creates an encrypted signature to retrieve the encrypted data in a data node. Second, it searches the hash table by using the upper n-bits of the key and generates the inverted list of partitions of the RP-tree. Third, the algorithm finds data nodes corresponding to each partition, in order to process the query by using the lower (b-n) bits of the key. Fourth, it determines whether or not the candidate nodes are included in the search area. Finally, it sequentially searches the signatures being stored in the data nodes for filtering out unnecessary candidates. To reduce I/O cost, we can load into a cache memory the signatures of the columns being frequently used.

4.2   Range Query and Partial-Match Query Processing

The range query processing algorithm is performed as follows. First, the algorithm finds the search range of the prefix-tree for a given query. Second, it selects the prefix sub-trees corresponding to the query and assigns them to threads to retrieve the data. Third, the exact-match query is performed on the assigned sub-trees by using multi threads. Finally, the algorithm determines whether or not the candidates retrieved from each thread are included in the query region. The final result is returned to the user.

To the best of our knowledge, there are no existing work to support a partial-match query on the encrypted data. Thus, we propose a partial-match query processing algorithm. The partial-match query processing algorithm is performed in the similar way to the range query processing one. First, the algorithm analyzes a query to find out a query pattern. Second, it stores the indexes of the prefix sub-trees corresponding to the query pattern. Third, the algorithm assigns them to the threads and retrieves candidate results. Finally, it merges the candidate results and returns the final result to the user.

4.3   View-Based Top-k Query Processing

Efficient top-k query processing algorithms have been an active research topic. The top-k queries aim to quickly retrieve k number of the highest ranking tuples in the presence of monotone ranking functions, which are defined on the attributes of underlying relations. For example, a user wants to find top 5 hotels where a cost combining both a distance from the user's current location and a payment per one day is the minimum. For a large-scale database, it is difficult to calculate the costs of all the data for answering a top-k query. To solve the problem, a Linear Programming Adaptation Algorithm (LPTA+) was proposed by Xie et al. [7]. LPTA+ selects views, which are the previous result set, similar to the given query. Then it calculates the costs of records stored in the selected views. Finally, it obtains the k number of results and returns them to the user.

However, the existing LPTA+ cannot process the top-k query on the encrypted database. For this, we propose a view-based top-k query processing algorithm over the encrypted data. First, in the preprocessing step, our top-k algorithm stores the top-k results with the original data. The cached results contain only the final scores being calculated with the given top-k queries, so it is difficult to obtain the original values from the cached results. Second, the algorithm filters out the unnecessary cached data by using a view selection technique used in LPTA+. Thirdly, the algorithm measures the similarity between the query and the selected view in three ways: (i) by counting the number of attributes being in both the query and the view, (ii) by measuring the angle between the view and the query, (iii) by measuring the distance between a unit-distanced point from the view and the query. The unit-distance point means one that is located in one unit distance apart from the starting point of the view. Finally, it transforms the score of the view using the similarity, and returns the k number of results whose transformed score are the smallest.

5 Performance Analysis

5.1 Experimental Environments

In this section, we present the experimental analysis of both our distributed index structure and our encrypted query processing algorithm. For our distributed index structure, we measure the degree of the data protection. In case of the exact-match and range queries, we compare our encrypted query processing algorithm with the existing CryptDB [4], in terms of the query processing time and the accuracy of the query results. In case of the top-k query, we compare our encrypted top-k query processing algorithm with the existing LPTA+ [7] because the CryptDB cannot support the top-k query on the encrypted data. In order to compute the similarity between data, our encrypted top-k query processing algorithm uses two types of similarity measurements, i.e., SVS and SVA. Our encrypted top-k algorithm using the SVS is named as ETkS whereas our top-k algorithm using the SVA is named as ETkA. For our experiments, we use an expanded Census dataset whose size is 2 GB. The dataset includes sensitive data such as name, married or single, number of children, sex, age, level of education, job, majority, salary, and income/expense by property. We did our performance analysis on the Window 7 Enterprise K system with Intel Core2 Quad 2.4 GHz and 2 GB memory.

5.2   Data Privacy of the Encrypted Record ID

For measuring the data privacy of our encrypted record IDs being generated by RP-tree, we compare a similarity between the original data and the encrypted record ID by using t-test statistics. The t-test is used to determine whether or not the means of two groups are statistically different from each other. First, the t-value is measured by Eq. (2), where $\bar{X}_i$, s_i, n_i mean the average of the data, the standard deviation of the data, and the number of the data in dataset i, respectively.

$$t = \frac{(\bar{X}_1 - \bar{X}_2)}{s(\bar{X}_1 - \bar{X}_2)}, \left(where \, s(\bar{X}_1 - \bar{X}_2) = \sqrt{\frac{s_1^2}{n_1} + \frac{s_2^2}{n_2}} \right) \tag{2}$$

Second, the t-test method measures a p-value that is used to quantify the statistical significance of evidence in the context of null hypothesis testing. If the p-value is equal to or smaller than the t-value, the observed data are inconsistent with the assumption that the null hypothesis is true, and thus the hypothesis must be rejected. Before calculating the p-value, a threshold value α, called the significance level of the test, is traditionally chosen as 5% or 1%. In our performance analysis, if the t-value is smaller than or equal to the p-value, the data privacy of our encrypted record ID is not preserved. Because the significance level is set to 5% in general, we measure the t-value and the p-value between two datasets with significance level = 5%. The t-value and the p-value are 34.45552 and 2.74E-72, respectively. Because t-value > p-value, the distribution of our encrypted record IDs is different from that of the original data. As a result, an attacker has a difficulty to obtain the original data from our encrypted record ID.

5.3   Exact-Match and Range Query Processing Time

Figure 4 shows the exact-match query processing time of our algorithm and that of the existing CryptDB over the encrypted data. Our algorithm shows 1.42, 1.5, 1.49, and 1.51 s with varying the size of the data from 500 MB to 2 GB, whereas CryptDB

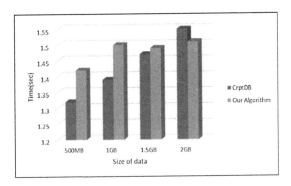

Fig. 4. Exact-match query processing time with varying the size of data

shows 1.32, 1.39, 1.47 and 1.55 s. Here, the incremental slope of the exact query processing time in CryptDB and our algorithm graph is 0.077 and 0.026, respectively. As a result, it is shown that our algorithm shows better performance than the existing work, when the size of data becomes larger than 2 GB. This is because our algorithm can select less number of candidates by using our prefix-tree based index structure.

Figure 5 shows the range query processing time for both our algorithm and the existing CryptDB, with varying the size of the query range from 0.0001% to 0.0005%. Here the size of the query range is computed as the ratio of the query range size over the data size. When the size of query range is equal to or smaller than 0.0002%, our algorithm is 1.1 times faster than CryptDB. However, when the query range size is greater than 0.0002%, our algorithm is about 15% slower than CryptDB. This is because the number of data with the same signature value is increased in our lower index as the query range size increases. Thus, the data transmission cost becomes greater than CryptDB when the query range size is larger than 0.0002%.

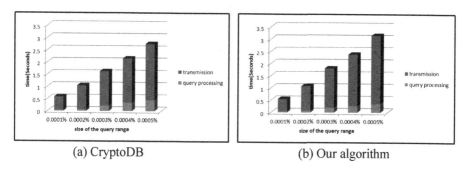

(a) CryptoDB   (b) Our algorithm

Fig. 5. Range query processing time with varying range size

5.4 Accuracy of the Top-K Query Result

Figure 6 shows the accuracy of the top-k query result of the LPTA+ algorithm and those of our encrypted top-k query processing algorithms (i.e., ETkS and ETkA), with varying the number of the cached query results. When the number of the cached results is 500, the accuracies of LPTA+, ETkS and ETkA are 65%, 70.4%, and 70.4%, respectively. This is because the LPTA+ makes use of all the cached query results to construct a candidate set. That is, some of the cached query results are irrelevant to the query, leading to low query result accuracy when the number of the cached results is large. Whereas, our encrypted top-k algorithms can filter out the cached results being not relevant to the given query, thus resulting in better performance on accuracy than LPTA+. In addition, the proposed Top-k query processing algorithm is useful because it can perform query processing on the encrypted database, unlike the existing LPTA+.

Figure 7 shows the query processing times of our encrypted top-k query processing algorithms. Because the LPTA+ cannot support top-k query over the encrypted data, we make experiments with both ETkS and ETkA when the number of cached query result is 500. Each cached result set contains both a top-100 query result with three attributes and the final score. Figure 7(a) shows the query processing time with varying

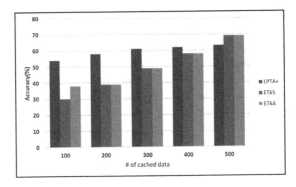

Fig. 6. Accuracy of the query results for the encrypted top-k query

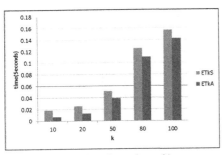

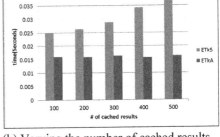

(a) Varying the value of k   (b) Varying the number of cached results

Fig. 7. Encrypted top-k query processing time

the value of k. When the k is 50, the query processing time of ETkS is 0.0515 whereas ETkA requires 0.0393 s. Because our ETkS needs to calculate the coordinates of the intersection points, its time complexity is $O(n2)$, whereas the time complexity of ETkA is $O(n)$. As a result, our ETkA shows better performance on query processing time than our ETkS. Figure 7(b) shows the query processing time with varying the number of the cached query results. For our experiments, we set the value of k as 20 and the number of the attributes as 2. When the number of the cached query results is 300, the query processing time of our ETkS is 0.0288 s, whereas our ETkA requires 0.0162 s. The reason is that the time complexity of our ETkS is $O(n2)$ due to calculating the coordinates of the intersection points.

6   Related Work

Because the existing query processing schemes decrypt an encrypted database to process a user's query, the original data may be revealed to a malicious attacker. Noiumkar and Chomsiri [8] confirms the lack of data encryption at rest on disk NoSQL databases. Encryption at rest is the term used to describe the encryption of the inactive data in a database. Similar technology is prevalent in the encryption of hard disks. NoSQL databases such as Cassandra [9] and MongoDB [10] provided their encryption schemes for their commercial product line. Meanwhile, query processing schemes over the encrypted database have been actively studied. They can be classified into two categories, SQL-like query processing scheme [4, 11] and aggregation query processing schemes [12–14].

Popa et al. proposed a typical SQL-like query processing scheme over the encrypted database, called CryptDB [4]. CryptDB encrypts data in a column-wise way by considering various query types, i.e., exact-match, range query and so on. To support the exact-match query, for example, it encrypts data using a deterministic encryption method such as AES. Poddar et al. proposed an encrypted query processing scheme on NoSQL database, named Arx [8]. Arx uses a similar infrastructure as the CryptDB, but uses AES as its strong encryption scheme. Meanwhile, typical aggregation query processing schemes over the encrypted database are as follows. First, Ge et al. [12] and Corena et al. [13] proposed a query processing scheme by using an additive homomorphic encryption scheme. The additive homomorphic encryption scheme has a property that the encrypted value of the summation of the original data is the same with the multiplication of their encrypted data. Especially, Ge et al. can reduce a computation cost by encrypting data in block units. Second, Thompson et al. [14] proposed PDSA (Privacy preserving Database As a Service). PDSA can aggregate data by using Shamir's secret sharing scheme on the distributed computing. The Shamir's secret sharing scheme divides a datum into several secret values and stores them in the different nodes. Here, a secret value is a constant term in an arbitrary polynomial equation. The scheme can calculate the real aggregation result by merging secret values without decrypting data.

The existing query processing schemes over the encrypted data have some problems. First, CryptDB cannot support operations among data with different columns because they use different types of encryption schemes depending on their attribute types. Second, the schemes proposed by Ge et al. [12] and Corena et al. [13] cannot support SQL-like queries, such as exact matching, range and join queries. In addition, the query processing cost is high because they use homomorphic encryption. Finally, the scheme proposed by Thompson et al. [14] may expose the original data when a service provider holds all secret values.

7 Conclusions and Future Work

Due to outsourcing a large amount of sensitive data, data encryption schemes to protect the sensitive data are required. Accordingly, it is necessary to develop not only a distributed index structure to efficiently manage the large-scale encrypted data, but also a query processing scheme over the encrypted data. For this, we proposed a distributed index structure that guarantees data privacy preservation and good performance for dealing with various types of queries. In addition, we proposed a secure query processing scheme that can process a query over the encrypted data without data decryption. Our query processing scheme provides both high query processing performance and high query result accuracy while preserving data privacy. We showed from our performance analysis that our proposed index structure and our query processing scheme are suitable for protecting large-scale sensitive data from attacker in data outsourcing environment.

As the future work, we plan to expand our work to handle elaborate types of queries, such as k-NN queries and skyline queries, over the encrypted data.

Acknowledgment. This work was partly supported by Institute for Information & communications Technology Promotion (IITP) grant funded by the Korea government (MSIP) (NO. R0113-16-0005, Development of an Unified Data Engineering Technology for Large-scale Transaction Processing and Real-time Complex Analytics). This research was supported by Basic Science Research Program through the National Research Foundation of Korea (NRF) funded by the Ministry of Education (grant number 2016R1D1A3B03935298).

References

1. https://research.fb.com/facebook-s-top-open-data-problems/
2. Advanced Encryption Standard (AES): NIST-Federal Information Processing Standards Publication 197 (2001)
3. RSA Laboratories.: RSAREF: A Cryptographic Toolkit Version 2.0 (1994). Homepage. https://www.rsa.com/en-us. Accessed 29 June 2017
4. Popa, R.A., Redfield, C.M.S., Zeldovich, N., Balakrishnan, H.: CryptDB: protecting confidentiality with encrypted query processing. In: Proceedings of the 23rd ACM Symposium on Operating Systems Principles, pp. 85–100. ACM (2011)
5. Dasgupta, S., Freund, Y.: Random projection trees and low dimensional manifolds. In: Proceedings of the Fortieth Annual ACM Symposium on Theory of Computing, pp. 537–546. ACM (2008)
6. Yazdani, N., Min, P.S.: Prefix trees: new efficient data structures for matching strings of different lengths. In: 2001 International Symposium on Database Engineering and Applications. IEEE (2001)
7. Xie, M., Lakshmanan, L.V., Wood, P.T.: Efficient top-k query answering using cached views. In: Proceedings of the 16th International Conference on Extending Database Technology, pp. 489–500. ACM (2013)
8. Noiumkar, P., Chomsiri, T.: A comparison the level of security on top 5 open source NoSQL databases. In: The 9th International Conference on Information Technology and Applications (ICITA) (2014)
9. http://cassandra.apache.org/
10. https://www.mongodb.com/
11. Poddar, R., Boelter, T., Popa, R.A.: Arx: a strongly encrypted database system. In: International Association for Cryptologic Research (IACR) Cryptology ePrint Archive (2016)
12. Ge, T., Zdonik, S.: Answering aggregation queries in a secure system model. In: Proceedings of the 33rd International Conference on Vary Large Data Bases, VLDB, pp. 519–530 (2007)
13. Corena, J., Ohtsuki, T.: Secure and fast aggregation of financial data in cloud based expense tracking applications. J. Netw. Syst. Manag. **20**(4), 534–560 (2012)
14. Thompson, B., Haber, S., Horne, W.G., Sander, T., Yao, D.: Privacy-preserving computation and verification of aggregate queries on outsourced databases. In: Goldberg, I., Atallah, M.J. (eds.) PETS 2009. LNCS, vol. 5672, pp. 185–201. Springer, Heidelberg (2009). https://doi.org/10.1007/978-3-642-03168-7_11

An Empirical Study on Performance Server Analysis and URL Phishing Prevention to Improve System Management Through Machine Learning

Antonio J. Tallón-Ballesteros[1]([⊠]), Simon James Fong[2],
and Raymond Kwok-Kay Wong[3]

[1] Department of Languages and Computer Systems,
University of Seville, Seville, Spain
atallon@us.es
[2] Department of Computer and Information Science,
University of Macau, Macau, China
[3] School of Computer Science and Engineering,
University of New South Wales, Kensington, NSW, Australia

Abstract. This paper tackles some important matters such as the server performance and the URL phishing. Nowadays the system management is a crucial issue and any potential failure needs to be detected quickly and, at the same time, to avoid URL phishing via defining rules in the firewall setting. An empirical study through data mining is conducted covering different prediction techniques. Lastly, some guidelines are provided to emit a critical view about what may happen and how to act immediately.

1 Introduction

Classification aims at predicting the class label for any forthcoming object and is an important topic within supervised machine learning [2]. Its single requirement is to have only samples which are labelled beforehand. The low storage cost has enabled to have data availability whenever and wherever. Additionally, a huge number of samples or properties may increase the computational cost to extract conclusive explanations [6]. The number of object types for a concrete problem may grow but this fact may happen from time to time where the period could be months, a few years or even several years. Therefore, techniques to reduce the instances [5] or the features [22] may be conveniently applied depending on the amount of data at a vertical or horizontal approach to drop information which refers to feature selection or instance selection, respectively. The goal of this paper is to analyse two cases of study from a predictive classification task perspective, to increase the models accuracy and if possible to try to accelerate the training process. One concerning the performance server to detect whether any failure is happening or in the other way round all is rightly operating and

© Springer Nature Switzerland AG 2019
M. Coppola et al. (Eds.): GECON 2018, LNCS 11113, pp. 199–207, 2019.
https://doi.org/10.1007/978-3-030-13342-9_17

another related to phishing websites especially to help the users to easily distinguish between a phishing or a legitimate URL (Uniform Resource Locator). The motivation of this research is to cover two problems: one more recent and trendy such as URL phishing and another not very new although very outstanding since servers are one of the pillars of any system management. The rest of this contribution is organised as follows. Section 2 describes briefly the problems to be faced in the context of supervised machine learning. Section 3 details the experimental process and the proposal concerning the data as well as the data pre-processing scenarios. Then, Sect. 4 depicts the results with a good number of classification algorithms. Lastly, Sect. 5 draws some remarks according to the empirical study and outlines some prospective works.

2   Problems

Nowadays the server is a core component in any organisation. Thus, any tentative failure needs to be detected as soon as possible or even to be suspected before its happening [18]. Another essential issue is that incoming data are flowing without stop. The server log or any type of software monitor or profiler is of utmost importance especially to track the server status at any moment [15]. It deserves also to mention the paper written by Hein et al. [8] which models a server system from many perspectives including the events and different kinds of failures. Data mining has been used to analyse the system dynamics as well as to infer some rules structured within a decision tree [11]. URL phishing is on the rise and is very frequent to hear about attacks to huge companies. There are some recent surveys about the topic [10] collecting some definitions [12] and the state-of-the-art from a theoretical point of view. An interesting definition published in [27] states "a phishing page as any web page that, without permission, alleges to act on behalf of a third party with the intention of confusing viewers into performing an action with which the viewer would only trust a true agent of the third party". Concerning the first scenario, the data from a server at Silicon Graphics Inc. are going to be analysed. The data were downloaded in 2014 from a repository located on Internet but unfortunately is no longer available and none study has been published up to now. The problem is very challenging and at the same time very well featured since there are measurements of properties belonging to different facets such as the CPU, the Input/Output, the disk, the net, the memory and the system calls. On the other hand, the website phishing data are publicly available at https://archive.ics.uci.edu/ml/datasets/phishing+websites which is included in the repository maintained by the University of California at Irvine. Moreover, there are many details in [13] which are very nice to provide any interested reader as a direct path to continue deepening about this topic during a while. Phistank directory is another interesting resource although a Natural Language Processing task is needed and hence is outside of our scope. Another way to distinguish URL phishing is through URL Ranking [4]. Simulation is gaining widespread acceptance in the current century especially to train in advance for any type of undesirable situation and the phishing is not the exception [1]. Heuristics is a very wide field to face Phishing [14]. The URL is very important to detect any kind of attack or hits through Internet [29]. Bayesian networks and support vectors are two interesting strategies to cope with URL Phishing [31]. Next section depicts some numerical values about the objects comprising both problems. It is not common at all to find contributions covering simultaneously the study of system management from a point of view of server performance in terms of failure analysis and phishing. The first part has some connections with [25]. Concerning the second part, the paper written by Ramanathan et al. [16] deals with phishing in an isolated way following a natural language processing approach.

3   Experimentation Setup and Proposal

It is crucial to describe how the experimentation is conducted. The experimental design follows a stratified hold-out [17] where two sets, called training and testing ones, containing three and one quarters, respectively, are created in order to first train a classifier and then to assess the performance leveraging the testing data. The stratification is a procedure to keep the original proportion or alike of samples for every class within the training and testing sets. Table 1 summarises the problems for the empirical study. Last row scores the average values. As can be seen the number of features is not higher than fifty although there are two well differentiated situations in the sense that in one case all the features are numeric and in the other all are nominal. As a practical rule of thumb, it is often convenient to apply feature selection from a number of attributes greater or equal than five in order to only include in the prediction model the most relevant attributes [24]. Feature selection is a wide field within data mining and there are many approaches depending on many factors [9]. This paper focuses on feature subset selection where the goal is to get a group of features that operate very accurately independently of the individual contribution of every single attribute taken one by one. Correlation measures are of particular interest to cope with numerical values. Alternatively, consistency-based metrics are very appropriate for nominal values and are unluckily not very common. We draft Correlation based Feature Selection (CFS) [7] to be applied on Server problem and CoNsistency-based feature Selection (CNS) [3] for Phishing. We have proposed very recently a CFS enhancement which applies CFS twice in a sequential fashion using different training data and the resulting methodology was called CHAracterisation of Features through Feature Subset Selection (ChaF2S2) [21]. CFS core is a heuristic to compute the worth of a subset of features which takes into account the usefulness of individual features to predict the class label along with the level of inter-correlation among them. An inconsistency scenario is defined as two instances having the same attribute values but different class labels. Since the full feature set always has the lowest inconsistency rate, FS thus attempts to minimise the number of features in the subset to reach a certain inconsistency rate. Figure 1 outlines the approach. Table 2 depicts the data preparation methodologies as well as the concrete techniques to be applied throughout this paper along with its parameters and values. As suggested by one of the reviewers, CNS has been also assessed in the context of ChaF2S2; the final situation is that combination of the first step and the second one with CNS retain all the features. For Phishing, the baseline scenario considers the full set of features and the enhanced situation is reached via CNS. CFS and ChaF2S2 through CFS twice are the baseline and enhanced scenarios for Server; for this case we do not test with the whole training set since for numerical features the application of CFS is more than recommendable. Table 3 details the feature space size for the baseline and enhanced situations for which next Section reports their results. As classification algorithms, we have chosen four classifiers, two out of them such as kNN (k Nearest Neighbours) and SVM (Support Vector Machines) [26] are very well-known and are even part of the top 10 algorithms within Data Mining [28]. The third candidate is BFTree (Best-first decision tree) [19] which has been successfully applied in some recent works [20]. Finally, the four algorithm is Random Forests (RF) which is an advanced classifier. Due to the fact that Phishing contains only nominal features, an extra classifier such as LBR (Lazy learning of Bayesian Rules) [30] is going to be applied to it. For kNN, the value of parameter k has been set to 1 and the method is going to be referred as 1NN. All the aforementioned classification algorithms fall in a different category in terms of the model representation such as rules, trees and so on.

Table 1. Problems

Problem	♯ Patterns		♯ Features				♯ Classes
	Training	Testing	Total	Numeric	Domain and type	Nominal	
Phishing	8291	2764	30	–	–	30	2
Server	720	239	44	44	Real (continuous)	–	2
Average	4505.5	1501.5	37				2

4   Results

This section reports the test results leveraging two performance measures such as the accuracy and Cohen's kappa [23]. The training time measured in seconds is also depicted for Phishing since the difference between the baseline and enhanced approach is the application or not of feature selection (FULL versus CNS) which affects clearly the computational cost for data sets with more than a couple thousands of training instances. For accuracy and Cohen's kappa higher is better which means that positive differences (Diff.) are improvements. For training time lower is better and therefore negative differences are enhancements. Equation 1 relates the time of the enhanced (CNS) and baseline scenarios (FULL) to compute the acceleration rate measured in percentage.

$$Acceleration\_Rate(\%) = \left(1 - \frac{time(CNS)}{time(FULL)}\right) 100 \qquad (1)$$

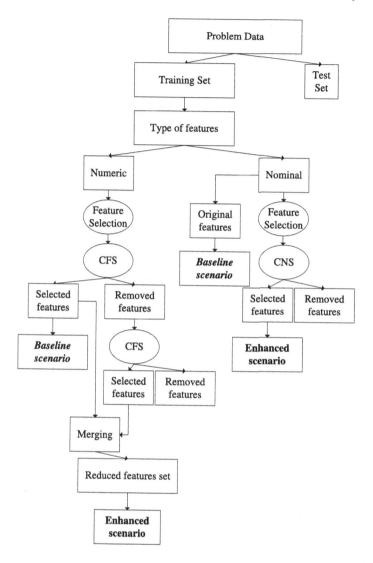

Fig. 1. Proposal.

Table 4 shows the test results for real-world problem Phishing. The best approach is 1NN without data preparation or any technique with feature selection excluding SVM, which are preferable to SVM. The selected features by CNS are as follow: having_IP_Address, URL_Length, Shortining_Service, having_At_Symbol, Prefix_Suffix, having_Sub_Domain, SSLfinal_State, Domain_registeration_length, HTTPS_token, Request_URL, URL_of_Anchor, Links_in_tags, SFH, Submitting_to_email, Redirect, age_of_domain, DNSRecord, web_traffic, Page_Rank, Google_Index, Links_pointing_to_page and Statistical_report. Classifier 1NN is very fast on this scenario though the worsening with

Table 2. Data pre-processing methodologies at feature level and parameter setup

Methodology	Method	Parameter	Value
Simple	CFS	Attribute evaluation measure	Correlation
		Search method	Best first
		Consecutive expanded nodes without improving	5
		Search direction	Forward
ChaF2S2	CFS + CFS	Attribute evaluation measure	Correlation
		Search method	Best first
		Consecutive expanded nodes without improving	5
		Search direction	Forward
Simple	CNS	Attribute evaluation measure	Consistency
		Search method	Best first
		Consecutive expanded nodes without improving	5
		Search direction	Forward

Table 3. Feature space of baseline and enhanced scenarios

Problem	Baseline		Enhanced	
	♯ Features	Method	♯ Features	Method
Phishing	30	None	22	CNS
Server	12	CFS	20	CFS + CFS

feature selection may compromise the scalability with a bigger sample size. The time for LBR plus Feature Selection is immediate; by its part BFTree requires above two seconds for about 8 thousands of training samples. SVM takes more than 6 s and does not seem very effective to detect website Phishing. Depending on the information volume it would deserve to process all the attributes or only the most relevant ones. Authors would like to remark some selected attributes given that four out of them (having_IP_Address, URL_Length, Shortining_Service and having_At_Symbol) are very easy to be taken into account in the daily navigation without sophisticated computation methods; another attribute as the age_of_domain could be detected from a software layer within the browser or in the queue of network traffic.

Table 5 exhibits the assessment results with the testing set on problem Server. The advanced method based on CFS, represented as CFS + CFS, is superior to the standard CFS in terms of accuracy and Cohen's kappa. BFTree reaches the performance peak followed by SVM and 1NN. The Cohen's kappa is always higher than 0.8 which means a strong confidence in the decision. BFTree takes

Table 4. Test results on problem Phishing

Classification algorithm	Accuracy			Cohen's kappa			Training time (s)			
	FULL	CNS	Diff.	FULL	CNS	Diff.	FULL	CNS	Diff.	Accel. (%)
1NN	97.21	96.89	−0.33	0.9435	0.9369	−0.0066	0.000	0.000	0.000	−
SVM	93.70	93.63	−0.07	0.8721	0.8707	−0.0014	6.828	6.281	−0.547	8.01
BFTree	95.04	95.08	0.04	0.8993	0.8999	0.0006	3.080	2.343	−0.737	23.93
LBR	93.78	94.50	0.72	0.8734	0.8885	0.0151	0.045	0.000	−0.045	100.00
RF	97.11	96.60	−0.51	0.9413	0.9309	−0.0103	0.164	0.094	−0.070	42.68

Table 5. Test results on problem Server

Classification algorithm	Accuracy			Cohen's kappa		
	CFS	CFS + CFS	Diff.	CFS	CFS + CFS	Diff.
1NN	88.70	91.21	2.51	0.7542	0.8078	0.0537
SVM	91.21	92.47	1.26	0.8048	0.8340	0.0292
BFTree	93.72	94.14	0.42	0.8634	0.8709	0.0075
RF	92.89	92.05	−0.84	0.8468	0.8261	−0.0207

the advantage of the advanced FS (ChaF2S2) and is able to beat even SVM. The selected features are from several typologies such as CPU (10 features, e.g. the average of busy CPU or idle CPU), I/O (6 features especially those tied with read or get), net (2 features) and memory (2 features: swap and flush). According to the followed methodology, the essential features cooperate with the important features in order to achieve a better prediction.

5 Conclusions and Further Works

This paper studied two real-world problems in the context of system management such as website phishing and the performance server in terms of failure happening, following a machine learning approach. Some data preparation methods based on standard and advanced feature selection computing different measures such as consistency and correlation properties, respectively have been shown to be particularly appropriate to cope with the aforementioned problems. Best-First Tree (BFTree) reported a good performance for both problems with some enhancements (more remarkable on Server data) in terms of predictive power provided by feature selection. Since the data sample of Phishing is medium, the data preparation method helps to build the prediction model in a shorter time. The more sophisticated feature selection is a key point for SVM when the features are not nominal (Server problem). For 1NN, the behaviour is alike to SVM. Finally, Lazy Bayesian Rule learning (LBR) exhibited a very accurate prediction on Phishing obtaining the classification model immediately.

As future works, we plan to extend the experimentation in terms of data preparation methods and classifiers. We also like to explore some alternative ways to detect URL phishing by means of browser plugins. Moving to server analysis, it is very convenient to seek approaches to log more properties to feature better the problem.

Acknowledgments. This work has been partially subsidised by TIN2014-55894-C2-R and TIN2017-88209-C2-R projects of the Spanish Inter-Ministerial Commission of Science and Technology (MICYT), FEDER funds and the P11-TIC-7528 project of the "Junta de Andalucía" (Spain).

References

1. Bahnsen, A.C., Torroledo, I., Camacho, L.D., Villegas, S.: DeepPhish: simulating malicious AI
2. Bzdok, D., Krzywinski, M., Altman, N.: Machine learning: supervised methods, SVM and kNN. Nature Methods (2018)
3. Dash, M., Liu, H., Motoda, H.: Consistency based feature selection. In: Terano, T., Liu, H., Chen, A.L.P. (eds.) PAKDD 2000. LNCS, vol. 1805, pp. 98–109. Springer, Heidelberg (2000). https://doi.org/10.1007/3-540-45571-X_12
4. Feroz, M.N., Mengel, S.: Phishing URL detection using URL ranking. In: 2015 IEEE International Congress on Big Data (BigData Congress), pp. 635–638. IEEE (2015)
5. Fung, G., Mangasarian, O.L.: Data selection for support vector machine classifiers. In: Proceedings of the sixth ACM SIGKDD International Conference on Knowledge Discovery and Data Mining, pp. 64–70. ACM (2000)
6. Giudici, P.: Applied Data Mining: Statistical Methods for Business and Industry. Wiley, Hoboken (2005)
7. Hall, M.A.: Correlation-based feature selection of discrete and numeric class machine learning (2000)
8. Heien, E., Kondo, D., Gainaru, A., LaPine, D., Kramer, B., Cappello, F.: Modeling and tolerating heterogeneous failures in large parallel systems. In: Proceedings of 2011 International Conference for High Performance Computing, Networking, Storage and Analysis, p. 45. ACM (2011)
9. Homenda, W., Pedrycz, W.: Pattern Recognition: A Quality of Data Perspective. Wiley, Hoboken (2018)
10. Khonji, M., Iraqi, Y., Jones, A.: Phishing detection: a literature survey. IEEE Commun. Surv. Tutor. **15**(4), 2091–2121 (2013)
11. Knobbe, A., Van der Wallen, D., Lewis, L.: Experiments with data mining in enterprise management. In: Proceedings of the Sixth IFIP/IEEE International Symposium on Integrated Network Management, 1999. Distributed Management for the Networked Millennium, pp. 353–366. IEEE (1999)
12. Lastdrager, E.E.H.: Achieving a consensual definition of phishing based on a systematic review of the literature. Crime Sci. **3**(1), 9 (2014)
13. Mohammad, R.M., Thabtah, F., McCluskey, L.: Predicting phishing websites based on self-structuring neural network. Neural Comput. Appl. **25**(2), 443–458 (2014)
14. Nguyen, L.A.T., To, B.L., Nguyen, H.K., Nguyen, M.H.: A novel approach for phishing detection using URL-based heuristic. In: 2014 International Conference on Computing, Management and Telecommunications (ComManTel), pp. 298–303. IEEE (2014)

15. Osterhage, W.W.: Computer performance optimization (2013)
16. Ramanathan, V., Wechsler, H.: phishGILLNET-phishing detection methodology using probabilistic latent semantic analysis, adaboost, and co-training. EURASIP J. Inf. Secur. **2012**(1), 1 (2012)
17. Refaeilzadeh, P., Tang, L., Liu, H.: Cross-validation. In: Liu, L., Özsu, M.T. (eds.) Encyclopedia of Database Systems, pp. 1–7. Springer, Boston (2016). https://doi. org/10.1007/978-0-387-39940-9
18. Schroeder, B., Gibson, G.: A large-scale study of failures in high-performance computing systems. IEEE Trans. Dependable Secur. Comput. **7**(4), 337–350 (2010)
19. Shi, H.: Best-first decision tree learning. Ph.D. thesis, The University of Waikato (2007)
20. Tallón-Ballesteros, A.J., Correia, L.: Medium and high-dimensionality attribute selection in Bayes-type classifiers. In: 2017 International Conference and Workshop on Bioinspired Intelligence (IWOBI), pp. 121–126. IEEE (2017)
21. Tallón-Ballesteros, A.J., Correia, L., Xue, B.: Featuring the attributes in supervised machine learning. In: de Cos Juez, F., et al. (eds.) HAIS 2018. LNCS, vol. 10870, pp. 350–362. Springer, Cham (2018). https://doi.org/10.1007/978-3-319-92639-1_29
22. Tallón-Ballesteros, A.J., Ibiza-Granados, A.: Simplifying pattern recognition problems via a scatter search algorithm. Int. J. Comput. Methods Eng. Sci. Mech. **17**(5–6), 315–321 (2016)
23. Tallón-Ballesteros, A.J., Riquelme, J.C.: Data mining methods applied to a digital forensics task for supervised machine learning. In: Muda, A.K., Choo, Y.-H., Abraham, A., N. Srihari, S. (eds.) Computational Intelligence in Digital Forensics: Forensic Investigation and Applications. SCI, vol. 555, pp. 413–428. Springer, Cham (2014). https://doi.org/10.1007/978-3-319-05885-6_17
24. Tallón-Ballesteros, A.J., Riquelme, J.C.: Low dimensionality or same subsets as a result of feature selection: an in-depth roadmap. In: Ferrández Vicente, J.M., Álvarez-Sánchez, J.R., de la Paz López, F., Toledo Moreo, J., Adeli, H. (eds.) IWINAC 2017. LNCS, vol. 10338, pp. 531–539. Springer, Cham (2017). https:// doi.org/10.1007/978-3-319-59773-7_54
25. Vargas, E., BluePrints, S.: High availability fundamentals. Sun Blueprints series, pp. 1–17 (2000)
26. Wang, J., Neskovic, P., Cooper, L.N.: Training data selection for support vector machines. In: Wang, L., Chen, K., Ong, Y.S. (eds.) ICNC 2005. LNCS, vol. 3610, pp. 554–564. Springer, Heidelberg (2005). https://doi.org/10.1007/11539087_71
27. Whittaker, C., Ryner, B., Nazif, M.: Large-scale automatic classification of phishing pages. In: NDSS, vol. 10, p. 2010 (2010)
28. Wu, X., et al.: Top 10 algorithms in data mining. Knowl. Inf. Syst. **14**(1), 1–37 (2008)
29. Zhang, J., Pan, Y., Wang, Z., Liu, B.: URL based gateway side phishing detection method, pp. 268–275. IEEE (2016)
30. Zheng, Z., Webb, G.I.: Lazy learning of Bayesian rules. Mach. Learn. **41**(1), 53–84 (2000)
31. Zouina, M., Outtaj, B.: A novel lightweight URL phishing detection system using SVM and similarity index. Hum. Centric Comput. Inf. Sci. **7**(1), 17 (2017)

Using Machine Learning for Identifying Ping Failure in Large Network Topology

Maged Helmy, Aurilla Aurelie Arntzen Bechina[✉],
and Arvid Siqveland

Department of Science and Industry Systems,
University of South-Eastern Norway, Kongsberg, Norway
aurillaa@usn.no

Abstract. It is well recognized in this digital world that, businesses, government, and people depend on reliable network infrastructure for all aspects of daily operations such as for i.e. Banking, retail, transportation and even socializing. Moreover, today, with the growing trend for the internet of thing, demands for a safe network management system has tremendously increased. Network failures are expensive: network downtime or outages should be avoided as it might affect business operations and might generate a tremendous cost due to the Mean Time to Repair in Network Infrastructure (MTR). This paper presents an ongoing work in exploring the use of machine learning algorithms for better diagnosis of network failure by using PING. To this end, we have analyzed 3 methods such Machine Learning (ML), Feature Selection with ML and hyperparameter tuning of ML. Within each method we used 3 algorithms such as KNN, Logistic Regression and Decision Tree algorithms and benchmarked them with each other's in order to define the best accuracy of ping failure identification.

Keywords: Machine learning · Network operation management · Ping failure

1 Introduction

With the increasing adoption of big data, cloud computing technology and a diverse category of interconnected devices (IoT), network topologies are growing even more complex. Therefore, it is important to prevent failure, downtime, reduce costs while increasing performance and security of the network. Network downtime or outage occurs regularly [1]. Today, many companies outsource their network management to the so-called Managed Service providers (MSP). These companies deliver networks, applications, systems and e-management services across a network to multiple enterprises, using a "pay as you go" pricing model. It is important for these MSP to provide reliable services to their customers. Our action-oriented research is based on a real case study that is provided by a Norwegian Managed Service Provider offering services in a network deployment, monitoring, management, and maintenance. Preliminary interviews with the network experts of this company indicated that the main challenges are due to Network Complexity and human factors. Their main concern is related to time consumption to identify the cause failure. Usually, network experts should analyze the

© Springer Nature Switzerland AG 2019
M. Coppola et al. (Eds.): GECON 2018, LNCS 11113, pp. 208–216, 2019.
https://doi.org/10.1007/978-3-030-13342-9_18

errors or events log manually. This paper presents a work in progress and focuses on identifying the best ways for network failure. They are many approaches to diagnosis network failure, in this research project we focus mostly on PING utility, as it is a requirement given by the case study. In this paper, we investigate if machine-learning algorithms could be used to better predict ping failure. To this end, we have defined a Knowledge Discovering Process (KDP) and compared several machine learning algorithms in order to determine the method that yield the best accuracy. This KDP is introduced in the next section. The third section presents methods and some results.

2   The Context of the Study

The Network structure can be very complex with each data point representing thousands of features. Identifying and fixing network issues can be structured in 4 steps:

- Validation of latency/issue
- Locating the area where the network is failing
- Identifying why it fails and provide means to remediate to the issue (Fig. 1).

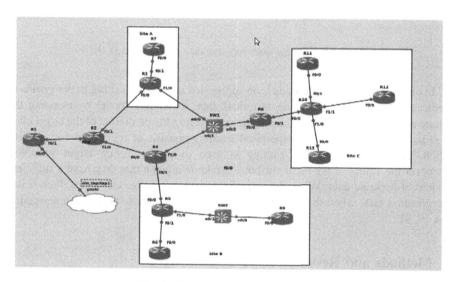

Fig. 1. Example of network topology

Using Ping Utility and traceroute can help to diagnose network failure and understand packet loss and high latency. Pinging several devices on the various network and comparing the round-trip times might give information about what part of the network is failing. The dataset is usually very large in a network topology with millions of properties for each data point, so and it can be considered as big data. Our dataset in our experiment consists of 2200 feature variables per row. Each row is a unique instance of the full network topology. The values are numerical and stored in the No SQL database. The data is captured using Zabbix which is a software for monitoring networks. Per day, this network topology can produce around 8 million data points for the topology shown. Based on literature review [2, 3], we have defined a specific approach to analyze the data such as Knowledge Discovery and Data Mining (KDDM) process (see Fig. 2). KDDM process can help to find non-trivial relations in the dataset and at the same time learn unbiasedly from historical data. Machine learning (ML) is a suitable tool to find patterns in data without having any prior knowledge of the dataset [4, 5]. There are three categories of ML [6]; supervised learning, unsupervised learning and reinforcement learning [7].

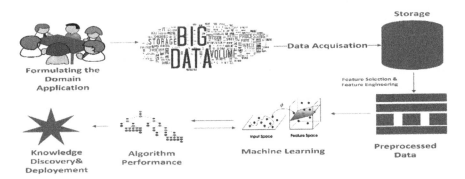

Fig. 2. Our knowledge discovery and data mining process (KDP)

In our study research, our focus is on supervised learning algorithm as our goal is to predict a value based on existing historical data. It builds a model by mapping the feature attributes to the target class and then comparing the prediction of the target class to ground truth. Furthermore, supervised learning consists of regression and classification. Moreover, supervised learning is used to predict future target variables, whenever the values of the input attributes are known given that they have a sufficient amount of accurate data. When the target class is a set of discrete values, then it is a classification task, when they are continuous numerical values, then it is a regression task [8, 9].

3 Methods and Results

We applied and compared three different exiting methods that are described below. For each method, we will compare with each other's three algorithms (KNN, Logistic Regression and Decision Tree) to determine which one will give the best accuracy of the ping status The common first step amongst the methodologies is to formulate the domain application. In this section, we present only few results for some algorithms. A more complete version of the work and results for each method and algorithm can be found [10].

3.1 Method I: Machine Learning Algorithms

ML algorithms are directly used to the preprocessed dataset to measure the performance. The algorithm's properties are adopted from the Scikit learn package [11]. It features various classification, regression and clustering algorithms and is designed to interoperate with the Python numerical. The default parameters will be used in the first step. In part 4.1 (Fig. 3), the K-Nearest Neighbors Algorithm (kNN) is classified as a lazy algorithm as it does not construct a general model from the dataset but rather classifies a single instance on every iteration [12]. Logistic Regression combines different weights with each variable to predict the outcome [13]. Decision Tree Algorithm finds a generalized set of rules to classify the instances in the dataset [14] (Table 1).

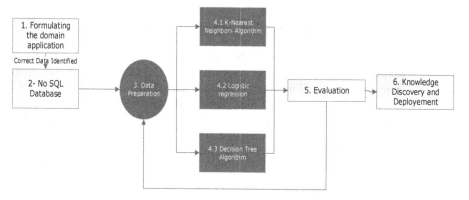

Fig. 3. Method I: machine learning algorithm

Table 1. Confusing matrix for KNN

Class target	Precision	Recall	F1-score	Support
0.0	0.8435	0.8291	0.8363	117
1.0	0.9430	0.9484	0.9457	349
Average	0.9180	0.9185	0.9182	466

Class target is set for the ping 1 or 0. Precision is defined as the proportion of positive identified correct. Recall means what proportion of the actual positives identified correctly. F1 Calculates harmony between 2 values. Implementation methods were programmed using Python language. Results of the first ML algorithm used is the kNN show that The K nearest neighbors can predict the network topology at an overall accuracy of 91.82% without any modification to the machine learning algorithm. The above table summarizes the result of the algorithm. For class zero, the precision score shows that the kNN labeled 84.35% of ping down correctly instead of ping up. The recall score shows that the algorithm can find 82.91% of all ping down in the dataset. The F1 Score shows the weighted harmony between the precision and recall and that is

83.63%. The total number of samples is 117, where 97 were labeled correctly, and 20 were labeled incorrectly. For Class 1, the precision score shows that the kNN labeled 94.30% of ping up correctly instead of ping down. The recall score shows that the algorithm can find 94.84% of all ping down in the dataset. The F1 Score shows the weighted harmony between the precision and recall and that is 94.5%. The total number of samples is 349, where 331 were labeled correctly, and 18 were labeled incorrectly.

We use the same approach for logistic regression and Decision tree algorithm. In conclusion, for the method I, the best accuracy is given by Decision Tree (95,7%) while kNN gave the least (91,82%) From the above, we can conclude that the decision tree was the most accurate and the kNN was the least accurate for classifying ping failures in network topologies.

3.2   Method II: Feature Selection with Machine Learning Algorithms

The feature selection is added as part of the data preparation stage, for example, it can remove features that provide redundant information. In Univariate Feature Selection: the best features are selected if they are higher than a certain threshold. This threshold can be tuned in three ways. *SelectKBest, Select Percentile* or *Generic Univariate selection*. In *SelectKBest*, the highest scoring features are defined by the user as value *K*. In *Select percentile*, the top *K* percentile feature is selected instead. In *generic univariate selection*, the feature selection method applies a configurable strategy to select the best feature (Fig. 4).

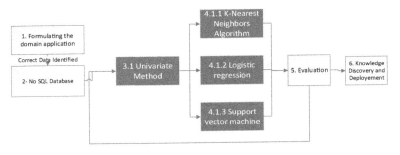

Fig. 4. Method II: feature selection with machine learning

In the second method, we applied feature selection to the architectural to analyze the classification accuracy of the algorithms and used two types of statistical methods, the chi2, and the mutual entropy. As an example, we will present the result performed on kNN. The same approach is used for Logistic regression and Decision Tree.

For the kNN algorithm, we can see that the overall algorithm has improved an overall accuracy from 91.86% to 96.99% when applying the chi2. The Fig. 5 on the left side shows the plot of accuracy against the number of features the chi algorithm has selected. The chi2 method has a range of optimal features from 28 features and a maximum of 80. The graph shows the features selected for accuracy on the test. The features selected are decided by the chi2 algorithm.

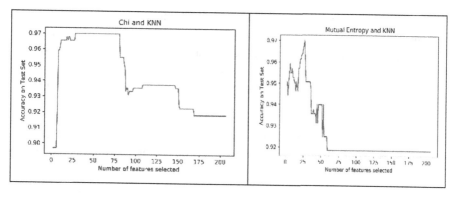

Fig. 5. Chi & KNN and mutual entropy & KNN

A more detailed breakdown of the algorithm prediction accuracy is shown in Table 2. For class zero, the precision score increased from 84.35% to 94.02%. The recall score increased from 82.91% to 94.02%. The F1 Score shows that the weighted harmony between the precision and recall increased from 83.63% to 94.02%. The total number of samples in class 0 is 117, from 20 labeled incorrectly to 7. For Class 1, the precision score shows that the kNN labeled 94.30% to 97.99% of ping up correctly instead of ping down. The recall score shows that the algorithm can find 94.84% to 97.99% of all ping down in the dataset. The F1 Score shows that the weighted harmony between the precision and recall increased from 94.5% to 97.99%. The total number of samples for class 1 is 349, where from 18 were labeled incorrectly down to 7 labeled incorrectly.

Table 2. kNN with CHI the tables.

Class target	Precision	Recall	F1-score	Support
0.0	0.8435 -> 0.9402	0.8291 -> 0.9402	0.8363 -> 0.9402	117
1.0	0.9430 -> 0.9799	0.9484 -> 0.9799	0.9457 -> 0.9799	349
Average	0.9180 -> 0.9700	0.9185 -> 0.9700	0.9182 -> 0.9700	466

Figure 5 shows application of Mutual entropy on kNN. The maximum accuracy and classification table report is the same as the mutual entropy which is at 96.99%. However, we can see that the number of features to be selected is different. The optimal number of features for maximum accuracy is 26. Therefore, we can conclude that chi2 and kNN are better combinations when it comes to predicting failures as it is more prone to noisy features in comparison to when applying mutual entropy.

We applied in a same manner chi2 and mutual entropy on logistic regression and decision tree and compared the results to see the improvement in accuracy prediction.

Table 3 shows that Chi2 as a feature selection to the network topology increases the kNN and Logistic regression accuracy labeling. The change in DT is non significant.

Table 3. Summary of results with univariate statistical method – CHI

Feature selection	Algorithm	No. features selected	Results
Univariate statistical tests (chi2)	K Nearest Neighbors	28–80	+5.18%
	Logistic Regression	72	+2.00%
	Decision Tree	22	+0.65%

Table 4 shows that the mutual entropy has a negative effect on the decision tree, reducing the accuracy slightly. Using feature selection does not necessarily increase the accuracy. Chi2 is better at feature selection than mutual entropy.

Table 4. Summary of results with mutual entropy

Feature selection	Algorithm	No. features selected	Results
Univariate statistical tests (Mutual Information)	K Nearest Neighbors	26	+5.18%
	Logistic Regression	18	+1.19%
	Decision Tree	76	−0.20%

3.3 Method III: Hyperparameter Tuning of Machine Learning Algorithms

Hyperparameter tuning is finding the best set of values to tune an algorithm to increase the accuracy of the prediction. There are two ways to tune the machine learning algorithm; either using the Grid Search Cross-Validation Method (CVM) or the Randomized Cross-Validation Method. In the grid search cross-validation method, all the combinations of parameters are generated and applied. The parameters that yield the highest accuracy with the correct classification is selected. The cross-validation prevents overfitting of the algorithm by having further split the training set. In the randomized search cross-validation, a set of random parameters are applied instead of all the combinations of parameters. This is faster and computationally less-expensive (Fig. 6).

The hyperparameter tuning is applied to find the optimal set of parameters for each algorithm.

Table 5 present the parameters of the ML algorithm to be tuned, with and without the optimal feature selection method. The same approach was adopted for Logistic regression and decision Tree.

Based on the results of the analysis of the three algorithms, we can conclude that the highest increase in accuracy to identify ping failures is achieved with Logistic Regression.

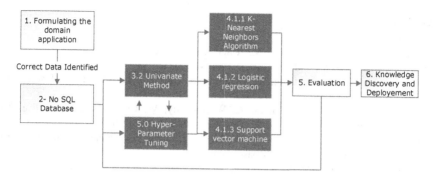

Fig. 6. Method III: hyperparameter with machine learning

Table 5. kNN with hyper-turned parameter and feature selection results

KNN parameters	Hyper-tuned kNN parameters	Hyper-tuned kNN parameters with feature selection
Number of neighbors	5 -> 3	5 -> 3
Weights	Uniform -> Distance	Uniform -> Distance
The algorithm to compute nearest neighbors	Auto	Auto
Leaf size	30 -> 10	30 -> 10
Power parameter value for Minkwoski distance	2 -> 1	2 -> 1
Change in accuracy	92% -> 96.85%	92% -> 97.24%

4   Conclusion

This paper presents and analyses the results of several methodologies applied for identifying ping failures in a network. Method 1 consisted of applying a machine-learning algorithm directly and consisted of the kNN algorithm, Logistic Regression, and the Decision Tree Algorithm. Method 2 was applying the univariate statistical method to reduce the number of irrelevant features passed to the algorithm to increase the accuracy of the classification. Method 3 was hyper-tuning the parameters of the machine learning ones with feature selection and ones without feature selection. The hyper-tuning parameter goal is to find a set of suitable parameters to increase the accuracy of the algorithm. The decision tree algorithm proved to yield the highest accuracy at method I. In method II, kNN had the highest increase in accuracy classification. In Method III, Logistic Regression with hyperparameter tuning and feature selection yielded the highest accuracy in predicting ping failures. The promising results from the test of the use case demonstrate that use of ML in improving identification of network failure has led to further directions in our work to explore the use of ML for network management. As future we intend to investigate the overfitting of the algorithms to the dataset. In addition, we will improve our algorithms in order understand why a certain host ping has failed.

References

1. Gill, P., Jain, N., Nagappan, N.: Understanding network failures in data centers: measurement, analysis, and implications, vol. 41 (2011)
2. Wu, X., Zhu, X., Wu, G., Ding, W.: Data mining with big data. IEEE Trans. Knowl. Data Eng. **26**, 97–107 (2014)
3. Sivarajah, U., Kamal, M.M., Irani, Z., Weerakkody, V.: Critical analysis of Big Data challenges and analytical methods. J. Bus. Res. **70**, 263–286 (2017)
4. Müller, A.C., Guido, S.: Introduction to Machine Learning with Python: A Guide for Data Scientists. O'Reilly Media Inc., Sebastopol (2016)
5. Brachman, R.J., Khabaza, T., Kloesgen, W., Piatetsky-Shapiro, G., Simoudis, E.: Mining business databases. Commun. ACM **39**, 42–48 (1996)
6. Robert, C.: Machine Learning, A Probabilistic Perspective. Taylor & Francis, Milton Park (2014)
7. Witten, I.H., Frank, E., Hall, M.A., Pal, C.J.: Data Mining: Practical Machine Learning Tools and Techniques. Morgan Kaufmann, San Francisco (2016)
8. Fawcett, T., Provost, F.: Adaptive fraud detection. Data Min. Knowl. Disc. **1**, 291–316 (1997)
9. Enke, D., Thawornwong, S.: The use of data mining and neural networks for forecasting stock market returns. Expert Syst. Appl. **29**, 927–940 (2005)
10. Helmy, M.: Identifying ping failures in large network topologies using machine learning, Master degree. University of South-Eastern of Norway, Kongsberg, Norway (2018)
11. Pedregosa, F., Varoquaux, G., Gramfort, A., Michel, V., Thirion, B., Grisel, O., et al.: Scikit-learn: machine learning in Python. J. Mach. Learn. Res. **12**, 2825–2830 (2011)
12. Zhang, M.-L., Zhou, Z.-H.: ML-KNN: a lazy learning approach to multi-label learning. Pattern Recogn. **40**, 2038–2048 (2007)
13. Glonek, G.F., McCullagh, P.: Multivariate logistic models. J. R. Stat. Soc. Ser. B (Methodol.), 533–546 (1995)
14. Quinlan, J.R.: Induction of decision trees. Mach. Learn. **1**, 81–106 (1986)

Special Topic Session - Blockchain Technologies and Economics

Smart Contracts for Container Based Video Conferencing Services: Architecture and Implementation

Sandi Gec[1], Dejan Lavbič[1], Marko Bajec[1], and Vlado Stankovski[2(✉)]

[1] Faculty of Computer and Information Science, University of Ljubljana,
Večna pot 113, Ljubljana, Slovenia
sandi.gec@fri.uni-lj.si
[2] Faculty of Civil and Geodetic Engineering, University of Ljubljana,
Jamova cesta 2, Ljubljana, Slovenia
vlado.stankovski@fgg.uni-lj.si

Abstract. Today, container-based virtualization is very popular due to the lightweight nature of containers and the ability to use them flexibly in various heterogeneously composed systems. This makes it possible to collaboratively develop services by sharing various types of resources, such as infrastructures, software and digitalized content. In this work, our home made video-conferencing (VC) system is used to study resource usage optimisation in business context. An application like this, does not provide monetization possibilities to all involved stakeholders including end users, cloud providers, software engineers and similar. Blockchain related technologies, such as Smart Contracts (SC) offer a possibility to address some of these needs. We introduce a novel architecture for monetization of added-value according to preferences of the stakeholders that participate in joint software service offers. The developed architecture facilitates use case scenarios of service and resource offers according to fixed and dynamic pricing schemes, fixed usage period, prepaid quota for flexible usage, division of income, consensual decisions among collaborative service providers, and constrained based usage of resources or services. Our container-based VC service, which is based on the Jitsi Meet Open Source software is used to demonstrate the proposed architecture and the benefits of the investigated use cases.

Keywords: Blockchain · Video-Conferencing · Container · Monetization · Smart Contracts

1 Introduction

Resources, such as computing infrastructures, Cloud services offers and software (Web servers, libraries and so on), combined together represent basis for the provisioning of high-quality software services. The various stakeholders usually need to collaborate in order to be able to produce such software services. This

© Springer Nature Switzerland AG 2019
M. Coppola et al. (Eds.): GECON 2018, LNCS 11113, pp. 219–233, 2019.
https://doi.org/10.1007/978-3-030-13342-9_19

particularly requires mechanisms for assuring monetization of the contributed added-value, transparency, security, trust, Quality of Service (QoS) assurance through Service Level Agreements (SLAs), and so on.

In order to facilitate flexible provisioning and consumption of resources and services, it is necessary to address various SLAs and monetization use cases of the stakeholders. In this study we concentrate on the analysis and implementation of various monetization approaches to collaboratively provide comprehensive software services to the end users. This includes SLA offers according to a fixed and variable pricing model, time-limited and quota-based provisioning, constraints-based access, revenue sharing, and other mechanisms making it possible to engineer and deliver software services with sufficient business flexibility. The goal of the present study is therefore to develop and evaluate an architecture that can be used to automate the process of software services provisioning and consumption.

Our approach relies on recent trends in the financial domain. Traditional currency transactions among people and companies are often facilitated with a central entity and controlled by a third party organizations such as banks. Bitcoin as a first decentralized digital currency was presented and launched as an alternative to centralized solutions. It is based on the Blockchain technology and has many benefits [1]. Through the adoption of Bitcoin among the world population the Blockchain technology presents opportunities in many areas, such as the Internet of Things, Cloud computing and software engineering in general. This has led to the launch of new dedicated cryptocurrencies – Ethereum[1] as a Smart Contracts (SC) ledger, IoTa[2] as an Internet of Things-based (IoT) ledger, Ripple[3] as a transaction cost optimization ledger, ledgers for anonymous transactions, protocols and other Blockchain systems covering a plethora of use cases. For example, in the Cloud domain the distributed storage system StorJ[4] is an alternative to commercial storage solutions such as Amazon S3 storage, Google Drive or Dropbox, and provides the same types of services with high encryption security and full transparency.

In this paper, we propose a new Cloud architecture for Container Image (CI) management containing a Web based Video-Conferencing (VC) application that assures high QoS and exploits the benefits of the Blockchain technology as a key component that differs from other existing solutions. Some key technologies, such as SCs, that are part of the architecture, are used to fulfill functional requirements, such as the needed agreements among the user and system roles. These are described in the following sections. By developing solutions for several generic monetization use cases, we aim to cover a potentially large number of stakeholders accordingly with their usage preferences, such as regular, occasional, demanding users and similar. This new architecture is designed having in mind the requirements for security, transparency and various monetization approaches

---

[1] https://www.ethereum.org/.
[2] https://iota.org/.
[3] https://ripple.com/.
[4] https://storj.io/.

through the use of SCs. The new architecture is implemented for one VC applications on top of Blockchain. This makes it possible to empirically compare it with traditional monetization approaches quantitatively and qualitatively.

The rest of the paper is organized as follows. Section 2 positions our work among other related works. Section 3 introduces Blockchain and Cloud computing in relation to the economy aspect. Section 4 presents and overview of the use cases based on using Blockchain. Section 5 presents the new architecture and its implementation. Section 6 presents a detailed overview of dynamic price monetization and explains the detailed workflow among the software services. Section 7 explains the relevance and significance of the obtained results, discusses the lessons learnt and provides some conclusions.

2   Related Work

In 1991 Haber et al. [6] presented a theoretical idea of the Blockchain concept, on how to certify digital documents in order to assure the tamper-proof data integrity. The first practical attempt of Blockchain technology was presented with the launch of the cryptocurrency Bitcoin [8] in 2009. Due to the Bitcoin simplicity of just sending and receiving digital assets, many researchers and Blockchain enthusiasts launched their own Blockchain cryptocurrencies. An interesting concept was presented by Buterin et al. [4] with the introduction of the Turing complete Smart Contracts (SCs) that can be compared to general (notary) contracts with limited, but at the same time sufficient functionalities that may cover several different use cases.

Blockchain monetization potentials were presented through practical applications and latest global trends by Swan [12]. The most potential applicable areas are digital assets registry management, solving the issue of billions of "unbanked" people, long-tail personalized economic services and payment channels in terms of creation of financial contracts executed over time. Yoo [14] analysed the potential of the Blockchain technology from the global financial perspective and thus also leaning on the use case of micro payments. A comparison between the current monetary system and Blockchain based is objectively presented by Ankenbrand et al. [2]. On the other hand Peter et al. [10] outlines the actual operational and regulatory challenges in terms of scalability, interoperability, standards, governance and others which should be defined with governments and financial institutes. All of these works explore ideas from a high-level perspective and do not provide any concrete monetization use cases, which are presented in our work.

Cloud computing has been dramatically adopted in all information technology (IT) environments for its efficiency, availability and hardware resource scalability. Therefore, Cloud computing architectures usually consist of variety of sub-systems, such as front-end and back-end platforms, Cloud based delivery and networks that are designed to address specific end user needs and requirements. Various requirements are also directed towards achieving high Quality of Service (QoS) and Quality of Experience (QoE). Some studies focus on the design of Cloud based orchestration systems that provide an intelligent delivery of CI based applications, e.g. a File Upload and a VC application as presented in our previous work [9]. This, however, is a first study that investigates the needs for monetization in the Cloud services economy which relies on the use of Blockchain. Our study aims to increase the overall system robustness in terms of system distribution, when various software services are developed, engineered, deployed and operated. Several highly focused studies have used Blockchain technology to improve product traceability and quality preservation, such as the study of Lu et al. [7]. The work of Xia et al. [13] focused on adoption of Blockchain for trust-less medical data sharing as a State-of-the-Art solution in the domain of medical Big Data. On the other hand the overall performance of such systems has not been addressed sufficiently and they have not been described in the context of SCs. An attempt of integration of SCs in an existing Cloud system for VMI and CI management was presented on H2020 ENTICE[5] project [5] as an agreement management component among users. The present work builds on top of the benefits of SCs, in order to define, develop and test various monetization strategies which may be useful in the domain of the Cloud services economy.

3   Background

Blockchain and SCs are technologies underpinning the design and implementation of our new architecture on top of which our Jitsi Meet[6] based VC system is provisioned. Blockchain (originally named block chain), is a continuously (time-based) growing list of records, called blocks, which are linked and secured using cryptography mechanisms (e.g. Bitcoin uses SHA-256). Each record ordinarily contains metadata – a reference (cryptographic hash) to the previous block, timestamp and transaction data. Blockchains are secure *by design* through broker-free (P2P-based) characteristics. Therefore, the alteration of the content in the blocks by single or even multiple node entities is very difficult to happen due to the high distribution rate. To summarize, key properties of the Blockchain are elevated distribution, no central authority, irreversibility, accessibility, timestamping and cryptography.

Smart Contract (SC) is a digital variation of a traditional contract which can be described as a protocol intended to digitally facilitate, verify and enforce the negotiation or performance by implementing arbitrary rules. In the design of our novel architecture, we rely on the Ethereum (see footnote 1) network with the built-in fully fledged Turing-complete programming language Solidity, which is used to design and implement SCs.

The life-cycle of a SC can be summarized in the following example. A developer designs a dedicated SC or uses an existing one. By doing that SC is considered as a template that is desired to be at least validated through specific tools. Each SC template may be deployed on a desired address to a production (mainnet) or development (testnet) network. Upon deployment the SC invokes

---

[5] http://www.entice-project.eu/.
[6] https://jitsi.org/.

the constructor only once, while the address owner usually has higher privileges, for example, it can destroy the deployed SC instance. Other participant, public or specific addresses, can trigger a SC instance through the supported functions that lead to sending of another transaction, triggering another SC, theoretically *ad infinitum* as illustrated in Fig. 1.

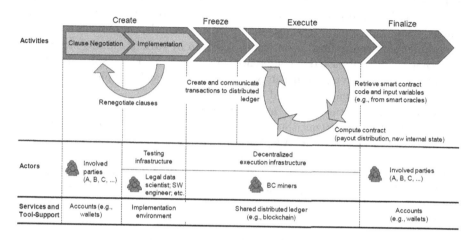

Fig. 1. The life cycle of a smart contract: phases, actors, and services [11].

In order to understand the monetization in the Cloud services domain we first need to know the economic ecosystem of the resources, services and environments deployed in the Clouds. In the first place, such environments must consider the total cost of ownership for the on-premises by considering the cost of the resources, equipment, computing and networking infrastructures, capital and the resources and services lifespan. All of these aspects can be described as the operational and maintenance costs of the services running on demand. Currently, there exist various special-purpose monetization services available, such as YouTube channel monetization, mobile applications advertising monetization and similar, which commonly have relatively expensive operational costs. It may therefore be possible to implement a flexible Blockchain monetization overlay on top of Cloud services.

In this paper, we focus on several use cases and requirements for monetization of a specific container-based VC service as an architectural overlay that may satisfy different types of end users. The developed architecture and its implementation is described in detail in the following section.

4 Video-Conferencing Application, Use Cases and Requirements Analysis

In this work we rely on advanced technologies, which are used in Cloud computing including Docker[7] for the management of containers, and Kubernetes[8] as a general purpose orchestration technology. Our VC application Jitsi Meet is therefore packed into Docker containers. In contrast to VMIs, containers can be spinned on and off much faster, practically within seconds, thus forming basis for fine-grained services orchestration, which is driven by various events (such as end users that need to run a video-conference). Whenever a specific software service is required, these services make it possible to dynamically deploy a container image and serve the particular end user or event.

Our goal was to study different monetization approaches that can be used by the stakeholders that contribute resources and services in order to develop a working VC software service. We present a basic comparison between conventional and Blockchain monetization on the Ethereum network.

In order to ensure feasibility of the monetization method applied in our system, we summarized basic properties of the Ethereum network and compared them with three popular payment systems: Visa, Mastercard and Paypal. The comparison results in Table 1 show that the overall cost of Ethereum monetization is significantly lower compared to traditional methods. Ethereum transaction cost represented with GWEI unit where $1\ ETH = 10^9\ GWEI$, is tightly dependent by the following factors: Ethereum network congestion, preferred transaction speed and the actual price of ETH coin. To minimize the ETH coin volatility there is a dedicated ERC20 token TrueUSD[9], which is an USD-backed, fully collateralized, transparent and legally protected. In addition to lower absolute transaction cost, SCs allows much more flexible transactions. For example, lock-in of transaction funds refers to the actual locking of funds in the SC till certain functional conditions has not been reached and releasing the funds, or just fractioned funds, to the appropriate user entity.

Table 1. Main properties of different monetization methods in the first half of the year 2018.

Monetization method	Transaction processing fee	Merchant or operational service cost [USD]	Lock-in of transaction funds
Visa	1.43%–2.4%	min 1.25%	Limited
Mastercard	1.55%–2.6%	min 1.25% + 0.05	Limited
PayPal	2.9%–4.4%	min 1.5%	Limited
Ethereum	1–40 GWEI[a]	0	Flexible

[a]https://kb.myetherwallet.com/gas/what-is-gas-ethereum.html

[7] https://www.docker.com/.
[8] https://kubernetes.io/.
[9] https://www.trusttoken.com/.

The monetization processes for our VC system from the viewpoint of the various stakeholders was described with several use cases as follows. The use cases describe different user needs based on actual usage patterns of resources and services. These can be achieved through the definitions of SCs. Another important aspect, which is considered concerns the income division among the parties, including the Cloud providers, container image deployment platforms, infrastructure providers and so on. In order words, we analysed the needs of all stakeholders who are in charge for an individual VC service setup processes. Following are seven identified monetization use cases that can be supported by Blockchain and Ethereum SCs.

Fixed price is the most basic definition of the use case. The VC system, depending on the QoS end user's requirements, offers the usage of VC service for a fixed price and fixed maximum period of time on demand. The agreement is reached when both parties signs the SC while the end user sends the required ETH funds.

Dynamic price, in addition to fixed price use case, offers higher flexibility of the actual VC service availability. For example, the end user knows the maximum time that he/she might need the VC service. After the agreement is reached the user pays the full price but the Ethereum funds are locked by the SC. The fund become unlocked if the maximum period is reached or if the end user stops using VC service by confirming the SC. In the unlocking phase the actual usage time is charged – the proportional Ethereum of unused time is returned to the end user while the rest is sent to the VC service.

Time-limited usage refers to the use case when the end user in advance agrees the per-minute conversation price, buys certain VC session minutes and uses the VC service gradually on demand. A typical usage is suitable for Big Brother-like TV shows and other content provided in real-time online.

Flexible usage period offers more flexibility for the end users that cannot define the exact period when the VC service will be used. In the proposed end user's period, the VC service has to be either available or the deployment of the VC service has to be optimized and thus very fast. The charge methodology follows the dynamic price use case and the SC contains also the minimum charge price for maintaining the fast deployment of the containerized VC service for the specific time period.

Division of income is a particular monetization process where the parties involved are those who enables the VC service hardware and software infrastructure. For example, one provider offers specific VC containers, another offers infrastructure and the third provider offers other services, such as monitoring. The division of income is agreed upon the parties in advance and validated through the SC. An overall advantage of this approach for the end user is reflected as a lower overall price of the VC service.

Consensus decision is an upgrade to division of income use case. The consensus is reached through a democratic voting SC, similar to the one proposed by Bragagnolo et al. [3]. As a result the management of VC service is divided among parties specialized for different HW/SW and service aspects and thus improved QoS/QoE, availability and overall cost reduction.

Constraint based focuses on the legal aspects determined by the end user's constraints. As an example EU General Data Protection Regulation (GDPR) compliant can be agreed by all end users through SC. In some cases geographical definition is also important and should be determined by the end user and VC system in advance (e.g. not all data services are allowed in specific countries). In general, by constraint limiting, either the price for the end user, either the potential QoS, changes (VC service price increase, decreased QoS) in the proposition of SC by the VC service.

In order to summarize, the rough analysis of the main properties of the monetization use cases addressing the various requirements are presented in Table 2.

5   Blockchain Based Architecture for Flexible Monetization of Cloud Resources and Software Services

The proposed architecture consist of various components that are packed as Docker CI. In order to address the proposed monetization use cases, we extend the existing VC architecture which includes an orchestrator [9]. While the focus of our previous work was the design of a Cloud based orchestration system that provides an intelligent delivery of two network intensive applications (File Upload and VC), the focus of the present work addresses the needs to facilitate flexible usage of the resources and software related to the provisioning of the VC service. We present an architecture to facilitate the VC application through the use of the Blockchain technology.

The novel architecture enables different monetization approaches among four pillar component layers: (i) Cloud providers, (ii) Implemented solution services, (iii) Blockchain components and (iv) Graphical User Interface (GUI). The implemented architecture is depicted in Fig. 2 and its main components are explained in the following.

Cloud providers are available on different geographical locations and are used for the deployment of VC service instances. The VC application components, which are packed into Docker containers can be deployed on demand when an agreement between the end user and the VC service provider is reached, and thus follow the SaaS delivery method. Besides, the deployed VC application instances can be monitored to provide metrics, such as usage time period and other QoS metrics that can be included in an SLA.

Implemented service solutions are composed of different components that comprehensively perform different tasks. For example, to estimate the potential QoS of the deployed VC service it is mandatory to have QoS metrics defined and

Table 2. Main properties of different monetization use cases.

Monetization use case	Minimum required SCs	SCs functions [description (SC)]	Roles involved
Fixed price	1	set price by owner, sign agreement	VC system, end user
Dynamic price	1	set price by owner, sign agreement, stop VC service	VC system, end user
Time-limited usage	$1 + n$ (n are number of accesses)	set price by owner (1), sign agreement (1), start VC session (n), stop VC session (n)	VC system, end user
Flexible period	1	set price by owner, sign agreement, stop VC service	VC system, end user
Division of income	2	set price by all VC enabling entities (1), set price by owner (2), sign agreement (2)	VC service enabling entities, end user
Consensus decision	2	set voters (1), vote by all VC enabling entities (1), set price by owner (2), sign agreement (2)	VC service enabling entities, end user
Constraint based	1	delegate constraints by owner, set price by owner, sign agreement, stop VC service	VC system, end user

measured, while the VC service is running. By doing this, the QoS metrics are modelled, while the Decision Maker identifies potential Cloud providers, where the container image should be started.

Blockchain components consist of a public Blockchain ledger, i.e. Ethereum and our developed SC templates supporting the proposed monetization use cases. See Sect. 4 for details. The distributed node infrastructure is maintained by the Ethereum community and offers a high level of distribution and availability. The main API technologies are written in the Java programming language, thus the core Ethereum bridge is the Java library ethereumj[10] embedded in the core layer API. The deployed SC instances may be triggered by end users and VC system. In all SC trigger executions and phases are updated by the actual content in the Ethereum blockchain network.

---

[10] https://github.com/ethereum/ethereumj.

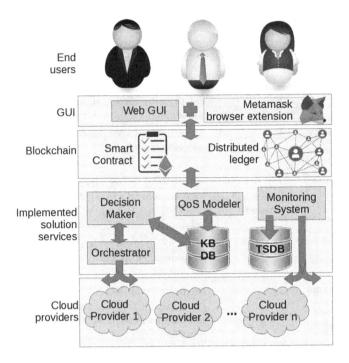

Fig. 2. High-level architecture of a SaaS VC use case.

Web based GUI is the main entry point for the end user. Each VC session consists of an end user subscriber and other participants. The subscriber in the Web GUI defines the preferences with QoS metrics and submits the request that is delegated to the backend component. According to the end user's preferences, the SC terms are proposed to the end user by our system. In all monetization use cases the agreement is reached when both parties signs the SC.

The general workflow of the architecture consist of the agreement among an end user and the VC service to determine the monetization use case, QoS user's requirements and the price. After the VC service deploys the CI application components, the end user gets unique URL for the VC session. The application life-cycle is completed when the end user finishes using the VC application or if any other condition occurs, e.g. the VC session time is exceed. In addition the resources are freed by the undeployment of VC session service on the Cloud providers' infrastructure. In the final stage the Ethereum SCs are signed by both parties, the end user and the VC service, and thus the final monetization flow is executed. A complete workflow with the dynamic price monetization use case is presented in the following section.

6   In-Depth Analysis of the Dynamic Price Monetization Use Case

In this section, we systematically describe the dynamic price monetization workflow, which is used by our VC software service. The selected use case is motivated by the transparent definition of the agreements among end user and VC system through an Ethereum SC.

The aim of the dynamic price monetization is increased flexibility for the end user, who cannot estimate the exact duration of the needed VC session. In comparison to traditional monetization methods (e.g. Paypal, Mastercard, Visa etc.), which primarily perform as a trustful entity and do lack transparency towards the end user, the SC Blockchain approach on the other hand is transparent, without central authority, and uses the defined functions in the Solidity programming language.

We integrated the SCs in our VC system in order to better understand the dynamic price monetization. The entities involved in the process are: (i) the end user who initiates the VC service, (ii) the Blockchain component which provides SC execution, (iii) the solution services, and (iv) Cloud providers that actually provide the running infrastructure for the VC application components.

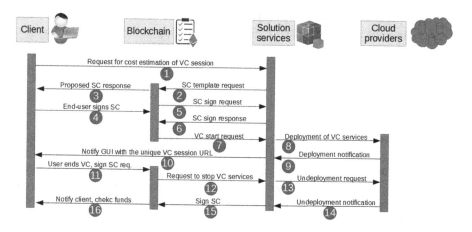

Fig. 3. Sequence diagram of the general VC use case.

The flow of interactions between the Blockchain component and the triggering entities is depicted in Fig. 3. In the beginning the end user sets basic properties through the Web GUI, for example the monetization type (dynamic price monetization), the maximum time period (one hour) and determines additional QoS preferences (availability 99.8 %, video quality high-definition).

The process flow is explained in the following.

1. Based on the end user's properties, the VC solution services estimate the price for the end user, based on QoS/QoE history trends, and deploy an SC instance,

2. The notification about the pricing policy is passed to the end user,
3. who approves the SC agreement by paying the full incentive, locked in the SC, for one hour of Ethereum funds (ETH) through the Ethereum bridge MetaMask[11], used as Web browser extension and monetization interface for ETH cryptocurrency.
4. The signed SC notification is passed to the VC solution services,
5. where it is signed by VC service address, and
6. further passed as a deployment request to the solution services, to be more concrete the QoS Modeler and the Decision Maker.
7. Further on, the deployment process is executed, and
8. on success,
9. the end user gets an unique Uniform Resource Locator (URL),
10 that can be shared among other VC participants.
11. Although it is a very common practice to share unique public URLs, this can be additionally enhanced, for example by registering individual participants addresses to obtain the VC session access. The VC session ends by the SC signature of the end user or the session automatically stops if one hour period is exceeded and no signing of SC is needed
12. In the final stage the undeployment of CI application instances is executed.
13–15. On success the final SC function, which unlocks ETH funds and executes proportional (time based) return of the ETHs or no return in case when time limit is exceeded. This step could be skipped but it is recommended as a final security check in case of any VC service anomalies (e.g. end user gets full refund in case when the availability of VC service dropped below 75%).
16. Finally, the end user gets the notification about the completion of the SC.

In order to outline the SC template for the dynamic price monetization, it is mandatory to define the key global attributes – lock time in seconds determined by current block timestamp, session start time, maximum price (refereed to lock time in seconds), end user address and VC owner address. Due to the usage of block based generation of the timestamp, the overall delay is inducted between the discrete time generation of blocks which is on average $15\,\mathrm{s}$[12].

An important aspect is the management of event triggers which are not supported in Ethereum SCs. A good practice recommendation is to either develop the service on our own or simply use Ethereum's Alarm Clock service[13], which is designed for SC event purposes.

A fundamental SC template that locks the ETH funds of the sender is described in Algorithm 1.

---

[11] https://metamask.io/.
[12] https://etherscan.io/chart/blocktime.
[13] http://www.ethereum-alarm-clock.com/.

Data: object payable {addressFrom, value}
Result: boolean success
if *basic conditions check* **then**
   | // in case the ETH value does not match the agreed price
   | return false;
else
   | // set the release time, address of the sender and deposit ETHs, while
   | the lockTimeSeconds is a global attribute
   | releaseTime = currentBlockTime + lockTimeSeconds;
   | balance = balance + value;
   | globalAddressFrom = addressFrom;
   | return true;
end

Algorithm 1: SC function for temporary locking of Ethereum funds.

Another pillar function is the VC service stopping notification that is triggered by the end user. After the successful trigger the ETH transaction to the VC service address is performed and another one to the address of the end user accordingly to the duration of the VC.

7 Discussion and Conclusion

The presented monetization approach in the previous section is overall less expensive compared to traditional monetization approaches. Beside the lower and relative fee policy not varying on the amount of funds to be sent, there are only initial integration costs and basically no further operational costs. Deployed and active SCs that temporary store the ETH funds of the end users, are accessible from the VC system APIs. However, the address containing the processed ETH funds owned by the VC system is stored in a cold storage wallet that make it secure in case of potential attacks.

This study presents a novel architecture that encapsulates Cloud principles and monetization possibilities through the usage of Blockchain. It is shown that SCs are suitable for establishing transparent mutual agreements among end users and VC system, and therefore facilitate the overall payment process without any additional cost for the service owners (e.g. bank, payment cards and other fees). Despite that, the specific SCs developed for the service needs to be carefully designed and validated (e.g. by using Oyente SC validation tool[14]), analysed and evaluated at different levels, such as code pattern comparison and simulation of actual usage with multiple parties on the same SC.

Following the implementation of our new architecture, we aim to investigate the QoS of the described monetization capabilities through performance measurements – actual transaction costs, block confirmation durations and other metrics related to the time metric.

---

[14] https://github.com/melonproject/oyente.

The Cloud domain with Blockchain technology poses some very specific research challenges for real-time applications (e.g. VC) that need to fulfil or enhance the functional requirements but at the same time introduce new functionalities, such as monetization. In our use case where the system methodology follows the pay-as-you-go concept, a software is deployed on end user's demand and SCs are used as an advanced agreement management tool among all stakeholders.

Acknowledgment. This project has received funding from the European Union's Horizon 2020 Research and Innovation Programme under Grant Agreement No. 815141 (DeCenter project: Decentralised technologies for orchestrated cloud-to-edge intelligence).

References

1. Anjum, A., Sporny, M., Sill, A.: Blockchain standards for compliance and trust. IEEE Cloud Comput. **4**(4), 84–90 (2017). https://doi.org/10.1109/MCC.2017. 3791019
2. Ankenbrand, T., Denis, B.: A structure for evaluating the potential of blockchain use cases in finance. Perspect. Innovations Econ. Bus. **17**(2), 77–94 (2018). https://doi.org/10.15208/pieb.2017.06
3. Bragagnolo, S., Rocha, H., Denker, M., Ducasse, S.: SmartInspect: solidity smart contract inspector. In: 2018 International Workshop on Blockchain Oriented Software Engineering (IWBOSE), pp. 9–18, March 2018. https://doi.org/10.1109/IWBOSE.2018.8327566
4. Buterin, V.: Ethereum white paper, updated 30 September 2015. https://github.com/ethereum/wiki/wiki/White-Paper. Accessed 30 Oct 2017
5. Gec, S., Paščinski, U., Stankovski, V.: Semantics for the Cloud: the ENTICE integrated environment and opportunities with smart contracts, January 2018. https://doi.org/10.5281/zenodo.1163999
6. Haber, S., Stornetta, W.S.: How to time-stamp a digital document. J. Cryptology **3**(2), 99–111 (1991). https://doi.org/10.1007/BF00196791
7. Lu, Q., Xu, X.: Adaptable blockchain-based systems: a case study for product traceability. IEEE Softw. **34**(6), 21–27 (2017). https://doi.org/10.1109/MS.2017. 4121227
8. Nakamoto, S.: Bitcoin: a peer-to-peer electronic cash system. http://bitcoin.org/bitcoin.pdf
9. Paščinski, U., Trnkoczy, J., Stankovski, V., Cigale, M., Gec, S.: Qos-aware orchestration of network intensive software utilities within software defined data centres. J. Grid Comput. **16**(1), 85–112 (2018). https://doi.org/10.1007/s10723-017-9415-1
10. Peter, H., Moser, A.: Blockchain-applications in banking & payment transactions: results of a survey. In: Nesleha, J., Plihal, T., Urbanovsky, K. (eds.) European Financial Systems 2017: Proceedings of the 14th International Scientific Conference, PT 2, Masaryk Univ, Fac Econ & Adm, Dept Finance; Inst Financial Market, pp. 141–149 (2017). 14th International Scientific Conference on European Financial Systems 2017, Brno, Czech Republic, 26–27 June 2017
11. Sillaber, C., Waltl, B.: Life cycle of smart contracts in blockchain ecosystems. Datenschutz und Datensicherheit - DuD **41**(8), 497–500 (2017). https://doi.org/10.1007/s11623-017-0819-7

12. Swan, M.: Anticipating the economic benefits of blockchain. Technol. Innovation Manag. Rev. **7**(10), 6–13 (2017). https://doi.org/10.22215/timreview/1109
13. Xia, Q., Sifah, E.B., Asamoah, K.O., Gao, J., Du, X., Guizani, M.: MeDShare: trust-less medical data sharing among cloud service providers via blockchain. IEEE Access **5**, 14757–14767 (2017). https://doi.org/10.1109/ACCESS.2017.2730843
14. Yoo, S.: Blockchain based financial case analysis and its implications. Asia Pac. J. Innovation Entrepreneurship **11**(3), 312–321 (2017). https://doi.org/10.1108/APJIE-12-2017-036

Deploying Blockchains for a New Paradigm of Media Experience

Georgios Palaiokrassas[(✉)], Antonios Litke, Georgios Fragkos,
Vasileios Papaefthymiou, and Theodora Varvarigou

National Technical University of Athens, Athens, Greece
geopal@mail.ntua.gr

Abstract. In this paper, we demonstrate the multiple points of innovation when combining multimedia content with blockchain technology. As of today, content creators (authors, photographers, radio and video reporters, data visualizers etc.) are publishing and sharing content (articles, photos, audio, video and combinations) on media/social networks but without the effective control over who is going to reuse this content. To this direction, we introduce a blockchain based service which successfully blends different technologies to provide more transparency on how the content is further tracked, promote openness, trust and security between participants, allow direct monetization for the content creator and expose the benefits of using blockchain technology as: a database of multimedia content, a novel payment method while using a dedicated created cryptocurrency, an insurance of proof of ownership through the exploitation of smart contracts running on Ethereum blockchain platform and a means to implement new solutions to value content based on quality. Additionally, by integrating Hyperledger Projects (Fabric, Composer, Explorer), we examine how their functionalities such as private and permissioned blockchain improve our system. Highlighting the importance of applications exploiting blockchain technology to efficiently support, store and retrieve data we utilized, integrated with blockchain and compared different solutions among which InterPlanetary File System (IPFS) and traditional databases such as MongoDB with GridFS tool.

Keywords: Smart contracts · Blockchain · Media content · Distributed applications

1 Introduction

In the recent years, there is a lot of interest on blockchain and distributed applications (Dapps), which explore the potential of blockchain technology. Naughton [1] suggests that blockchain technology could be 'the most important IT invention of our age', while Mougayar [2] that it is 'at the same level as the World Wide Web in terms of importance'. Blockchain-based technologies have the potential to resolve some of the current challenges of Media Industry, which have been heavily affected by the ubiquitous availability of content, by offering solutions for micropayment-based pricing models, monetization, copyright infringements and privacy [3]. Blockchain allows transactions to be peer-to-peer without the involvement of a Trusted Third Party

© Springer Nature Switzerland AG 2019
M. Coppola et al. (Eds.): GECON 2018, LNCS 11113, pp. 234–242, 2019.
https://doi.org/10.1007/978-3-030-13342-9_20

(TPP) and according to [4] it has two major features that make it highly attractive as a store of data. Firstly, data written to it is immutable and therefore provides an auditable record of events that cannot be modified. Secondly, an exact copy is maintained in a large number of independent locations and consequently, there is no central point of failure. Ethereum, the second-largest blockchain network by market capitalization, was the first platform to introduce the concept of a smart contract that could be deployed and executed in a distributed blockchain network. The Ethereum protocol is public so the terms of each contract can be viewed by anyone accessing the Ethereum blockchain network. Recognizing the potential of smart contracts on the Ethereum blockchain, Bogner et al. [5] examined the use of a Dapp for the sharing of everyday objects, achieving data privacy and avoiding users' repetitive sign up for each platform. In a similar direction, a trial rights management system for videos has also been implemented [6], using blockchain technology and trying to couple videos with the rights information and to include them in the transaction issued by the licensor.

The rest of the paper is structured as follows: Sect. 2 provides an overview of our proposed service, while Sect. 3 presents the implementation details and the developed smart contracts. Section 4 gives experimental results after running the service.

2 Use Case Overview and System Components

This paper introduces a distributed application that blends several technologies including smart contracts deployed on the blockchain, web services and databases to produce a service that offers solutions to issues such as compensation of content contributors, personalization, direct connection of content producers with consumers and digital rights management. The users who could benefit most from the presented system are on the one side artists, who would expose their content online, reach large audiences and have income directly without any intermediaries according to their specified terms. On the other side end users such as advertisers, news sites, publishers will have access to a big, constantly updated collection of original content generated by users, and immediately use it for their purposes after compensating the creator, without any other concern regarding digital copyrights. It should be noted that the proposed system does not handle the topics of checking the originality of the content and the verification of ownership of the specific content. Such topics are outside of the current study scope. The proposed system acts as a widely used decentralized repository where content providers (professional or not) can monetize their content. An indicative use case includes a photographer or any person capturing a significant event (e.g. the rescue of citizens after a big disaster) using his camera/smartphone, who ensures he is the first person taking this photo by inserting into the blockchain and receiving the transaction hash/timestamp. At the same time, various big media outlets (such as news agencies, online newspapers or news sites) who might want to officially use this user-generated content can now, through the proposed system, compensate the content provider by paying with cryptocurrency. To accomplish this, our deployed contracts communicate with each other, verify that the buyer has sufficient funds and perform the payment following the billing schema. Media-Tokens (our specifically created cryptocurrency) are transferred from consumer's account to creator's account and they both receive the transaction hash, as a proof of this deal. In the background, the Billing Manager is running, communicating with other contracts and external resources to check the views of each item and the current timestamp and send Tokens. Our proposed system consists of the following components (Fig. 1):

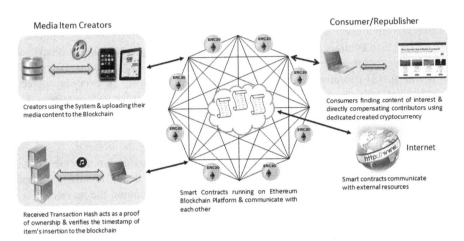

Fig. 1. Overview of our system: the media item creator sets specific rules and uploads content on the blockchain. The consumer further exploits it, after agreeing to the specified rules and the billing schema. The contracts automatically communicate with each other and with external resources to register the content, ensure its ownership and execute the necessary payments.

1. Web Application: The instances of the web application handle the interactions among the users and the system providing GUIs (Fig. 2) and interacting with the smart contracts deployed on the Ethereum blockchain, employing Web3 Javascript API. We have used Truffle development framework [http://truffleframework.com/docs] to build and deploy our contracts to the Ethereum network and HTML5/Javascript to develop the different interfaces. Two main interfaces are provided: (i) Content Creator: the creator using the dedicated user interface registers to the System. After specifying the details, he uploads the content to the blockchain; (ii) Content Consumer: the consumer is able to view all the stored media items in the blockchain, search among them using specific keywords and retrieve them (including the related information provided by the creator). Subsequently using the interface, he could consume the desired content and directly compensate the contributor.

2. Smart Contracts deployed on Ethereum Blockchain Application Platform: Ethereum is a decentralized blockchain platform that runs smart contracts, stored publicly on every node of the blockchain and every node is calculating all the smart contracts. We have developed three smart contracts and the functionality of communication among them as well. They have been implemented in Solidity programming language [https://solidity.readthedocs.io/en/develop], a contract-oriented, high-level language whose syntax, similar to Javascript. It supports inheritance, libraries, complex-defined types and is designed to target the Ethereum Virtual Machine.

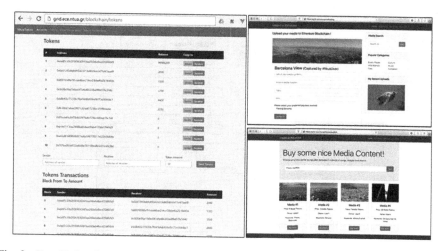

Fig. 2. Top-Right: the creator logs in the application and uploads a new media item to the Blockchain after providing the necessary information; Bottom-Right: The Content Consumer connects to the system and has access to all the uploaded media items; Left: User Interface of the Media-Token

3   Implementation Details

We implemented the functionality of communication among contracts, as well as with external resources (Fig. 3). For example, when a user wishes to exploit a media item through the user interface, which directly communicates with the Media Contract, the latter in the background communicates with the Token Contract to verify that this user has sufficient funds. Afterwards, the two contracts communicate again to perform the payment and exchange the transaction hashes, so the user has a proof of the payment. The Billing Manager Contract also communicates with the two other contracts, in order to perform the payments according to the chosen billing schema.

3.1   Media Content Smart Contract – "Blockchain as a Database of Content"

Media Content Contract handles the interaction between the content creator and our proposed system. The content creator logs in the system using the dedicated GUI provided by the web-app. He then chooses the new media item (article, photo, audio, video or other) to be uploaded to the blockchain and specifies the related information. In this direction, we have developed an HTML5 page where the user completes all the parameters which are then stored in the smart contract. The parameters stored are: URI of the stored media content; price in Tokens and billing schema, related tags, keywords other related commends and information. This smart contract supports numerous functions, e.g. the upload_media function and when a new item is uploaded an Event called Uploading is produced. The arguments of that event are the address of the creator and the tags of the uploaded media. All these arguments are stored in the

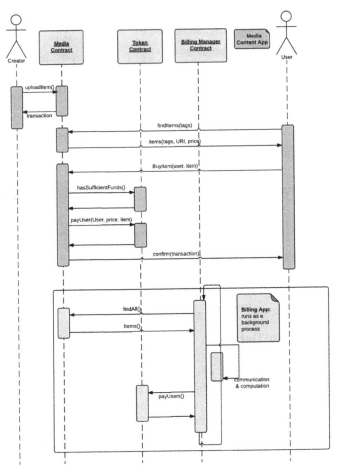

Fig. 3. Left: Sequence diagram of users' interaction with the distributed application deployed on the Ethereum blockchain and the communication among the three smart contracts.

specific transaction log, a special data structure in the Ethereum blockchain. After triggering this event, a series of Javascript callbacks in the Dapp take place and in the end the media content creator receives a transaction hash, which corresponds to a specific transaction, block and timestamp in the Ethereum blockchain. Recognizing the importance for an application exploiting blockchain technology to efficiently support, store and retrieve data we utilized and integrated with blockchain different solutions:

1. Content creators could store the digital content in their own database or alternatively use our system's database, since it is not efficient to store digital content (e.g. video) in the blockchain. So, providing their own URI to the blockchain, which is a link to where the actual data is stored, they retain exclusive control over their contents or they could use our MongoDB as a storage for their data. Media content storage in our database is achieved utilizing GridFS tool which is the MongoDB specification

for storing and retrieving files, such as images, music, videos, that exceed the BSON-document size limit of 16 MB. This approach has proven to be efficient with high throughput in terms of storing and retrieving media content of any type. Having a service, which also administrates the database, allows us to ensure that only users who have paid are able to receive the content. However, we could not neglect the centralized nature of this modeling having a traditional database.

2. IPFS: InterPlanetary File System. Aiming to a fully decentralized application we integrated blockchain with IPFS, a peer-to-peer version-controlled protocol and filesystem, run by multiple nodes, storing files submitted to it [7]. It combines distributed Hash Tables, Block Exchanges and Merkle Trees. Users upload content to IPFS and place its unique hash code (address of the file) to the blockchain.

3.2 Media-Token Smart Contract

In the recent years, many digital cryptocurrencies have been introduced, as also presented on a technical survey on digital cryptocurrencies [8]. For our system and for research purposes we developed a digital cryptocurrency named "Media-Token", in the form of a smart contract running on Ethereum public test network. It is fully compliant with ERC20 token standard [9], describing the functions and events an Ethereum contract has to implement. ERC20 enhances and ensures the interoperability among different smart contracts and distributed applications deployed on Ethereum blockchain platform. In addition, we implemented a dedicated user interface using HTML5 and Javascript libraries (including Web3.js) which allows us to monitor all the transactions of our Media Token and all the balances of all the accounts (Fig. 2).

3.3 Billing Manager Smart Contract

Our last deployed contract constantly monitors the other smart contracts, handles the communication with external sources and is responsible for executing payments according to the billing schema, chosen by the content creator while initially uploading the media items. The aforementioned smart contracts offer a dynamic billing mechanism with three options: (i) a fixed amount of Tokens is instantly transferred to the creator's account when the content is consumed; (ii) an amount is transferred to the creator's account once every day (or week/month) for the specified coming period; (iii) a dynamic schema where some Tokens are initially transferred and a fixed amount gets transferred when some conditions are met (e.g. the creator initially receives 1 Token and 1 Token/1000 views). This contract communicates with other contracts deployed on the blockchain and with external sources in order to receive valuable real-world information for the application such the total views of each item deployed on the blockchain. We used a solution based on Oraclize [https://github.com/oraclize], enabling a function to connect to a URL and retrieve information such as the views of each media item (which are exposed in a predefined format). This contract is regularly triggered in order to communicate with external sources and to perform the payment at specific timestamps and if some conditions are met e.g. every day in the morning (Fig. 4).

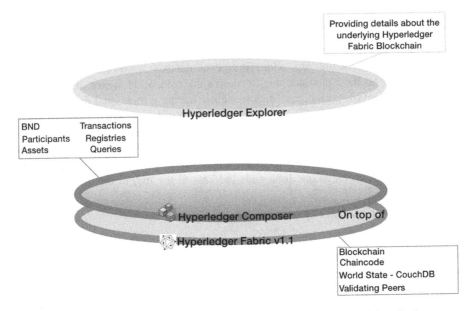

Fig. 4. Overview of the overall system architecture based on Hyperledger Projects

3.4   Deploying Hyperledger Platform and Projects

Another version of the aforementioned service was implemented based on permissioned blockchains of Hyperledger Projects offering more safety features such as identity management with Membership Service Providers, faster transactions and less computational expensive consensus algorithms compared to PoW. We should note that by granting open access to all users, we would have a public-like blockchain. Our solution was based on the following Hyperledger open source platforms: (i) Hyperledger Fabric offering the permissioned blockchain for the users of our system. Since it has no built-in cryptocurrency, we defined a digital token as an asset in the Composer's model file (.cto). Additionally, we define the media asset, visible to all granted users who are defined in the Access Control File (.acl); (ii) Hyperledger Composer facilitating the smart contracts ("chaincode") development and the definition of the business model. (iii) Hyperledger Explorer, a powerful tool providing details related to a channel, specific block information, peer list, chaincode list and transactions. The Media Content Creator, who is a granted blockchain user, creates a new media content asset (defined in the model file) by specifying its details, which is then uploaded and saved in the blockchain in an encrypted ASCII string format using the Base64 scheme. Afterward, a buyer can check all the available media content assets stored in the Fabric blockchain, search among them by using specific tags. Finally, the transaction logic, which is defined in the Script File of Hyperledger Composer's runtime, is responsible for checking whether the balance of the buyer suffices for the corresponding Media Content asset. If so, after the buyer's tokens are transferred to the creator's balance, the actual content is decoded according to the Base64 scheme and is saved on the buyer's local file system. The aforementioned transaction that took place between the two users is recorded in the Fabric blockchain and is visible from the Explorer's GUI.

4 Running the System

When the Creator uploads a new media item, a transaction takes place. Every 15 s a new block is added to the blockchain with the latest transactions according to Ethereum's mining process based on PoW algorithm which leverages EtHash [https://www.ethereum.org/ether]. Every time a transaction is created in Ethereum, the Transaction Hash is returned, helping to keep track of a transaction in progress or specifying the block where this transaction is included in and the related timestamp. The latter serves as a proof of ownership, ensuring that this media item belongs to this user and was inserted in the blockchain at a specific timestamp. When the consumer logs into the dedicated web-interface, he searches for the media item he wishes to exploit and sends funds to the creator through a series of functions and contracts communications. Content transactions will be secured using micropayments, made with Media-Tokens, in conjunction with the Ethereum ecsign and ecrecover functions. Which implement the elliptic curve digital sign algorithm, authenticating the addresses/users in the blockchain, refusing access to users who try to impersonate creators or consumers. The transfer of Tokens is also visible in Ethereum Wallets, which watch the specific contracts and act as a gateway to Dapps on the Ethereum blockchain and allow to hold and secure Ethers and other crypto-assets built on Ethereum, as well as write, deploy, use and watch smart contracts. Hyperledger implementation on the other side provides more potential for queries, since it supports a CouchDB database and allows the definition of different business models using channels.

Regarding the data storage, by running the system with MongoDB deployed, we benefited from the high throughput of a document-oriented central DB. However, this centralization does not coincide with the decentralized nature which Ethereum advocates leading to a Single Point of Failure (SPOF) of the whole application. Facing the aforementioned drawback, IPFS being a peer-to-peer (p2p) filesharing system and Blockchain's complementary component, settled exceptionally the SPOF problem, furnishing low latency and data distribution (Table 1).

Table 1. Comparison of the two different implementations of our system

	Implementation based on Ethereum	Implementation based on Hyperledger
Performance/Throughput	~15 TPS	<3500 TPS
Access	Open read/write	Permissioned read and/or write
Security	Proof of work	Pre-approved participants
Identities	Public keys as identities	Known/Granted identities

5 Conclusions

The current work presented in this paper, successfully combined blockchain with web technologies and user-generated content. In the future, the authors plan to expand the service for different use cases such as for music industry digital content broadcasting in open, blockchain enabled markets and media content delivery through WebTV.

Acknowledgment. The research leading to these results has received funding from the European Commission under the H2020 Programme's project Bloomen (grant agreement nr. 762091).

References

1. Naughton, J.: Is Blockchain the most important IT invention of our age? The Guardian (2016)
2. Mougayar, W., Vitalik, B.: The Business Blockchain: Promise, Practice and Application of the Next Internet Technology. Wiley, Hoboken (2016)
3. Deloitte: Blockchain @ Media, A new Game Changer for the Media Industry (2017)
4. O'Dair, M., Beaven, C.M., et al.: Music on the blockchain. Middlesex University (2016)
5. Bogner, A., Chanson, M., Meeuw, A.: A decentralised sharing app running a smart contract on the ethereum blockchain. In: Proceedings of the 6th International Conference on the Internet of Things, pp. 177–178 (2016)
6. Fujimura, S., et al.: BRIGHT: a concept for a decentralized rights management system based on blockchain. In: 2015 IEEE 5th International Conference on Consumer Electronics-Berlin (2015)
7. Chen, Y., et al.: An improved P2P file system scheme based on IPFS and Blockchain. In: IEEE International Conference on Big Data (Big Data) (2017)
8. Tschorsch, F., Scheuermann, B.: Bitcoin and beyond: a technical survey on decentralized digital currencies. IEEE Commun. Surv, Tutor (2016)
9. ERC20. http://theethereum.wiki/w/index.php/ERC20_Token_Standard

Is Arbitrage Possible in the Bitcoin Market? (Work-In-Progress Paper)

Stefano Bistarelli, Alessandra Cretarola, Gianna Figà-Talamanca, Ivan Mercanti$^{(\boxtimes)}$, and Marco Patacca

Università di Perugia, Perugia, Italy
{stefano.bistarelli,alessandra.cretarola,gianna.figatalamanca,
ivan.mercanti,marco.patacca}@unipg.it
http://www.unipg.it

Abstract. Bitcoin is a digital currency traded on different exchanges for different prices; this feature implies important issues about arbitrage opportunities. In this paper we investigate whether strong or weak form of arbitrage strategies are indeed possible by trading across different Bitcoin Exchanges. Our investigation, both theoretically and practically, gives as a result that arbitrage is indeed possible.

Keywords: Bitcoin · Application · Arbitrage

1 Introduction

The white-paper on Bitcoin appeared in November 2008 [8], written by a computer scientist using the pseudonym "Satoshi Nakamoto". His invention is an open-source, peer-to-peer digital currency, named Bitcoin. Transactions in Bitcoin do not require a central authority, neither a traditional financial-institution involved as an intermediary. The key innovation in Bitcoin is its decentralized core technologies [6]. This system allows to have an independent currency, not subject to the control of central authority and without inflation. Recent literature claims Bitcoin as a volatile stock rather than a currency, [10]. An important issue about Bitcoin prices is that it is traded on different web-exchanges for different prices hence it does not obey to the usual law of unique price. On the web there are also some sites that compare in real time the price of Bitcoin in different exchange as well as the price history in order to decide what are the best exchanges for buying and for selling Bitcoins at a fixed point in time. The most

M. Patacca—This research is supported by project "REMIX" (funded by Banca d'Italia and Fondazione Cassa di Risparmio di Perugia) and projects "ComPAArg" and "Argumentation 360" (funded by "Ricerca di Base 2015–2016", University of Perugia).

© Springer Nature Switzerland AG 2019
M. Coppola et al. (Eds.): GECON 2018, LNCS 11113, pp. 243–251, 2019.
https://doi.org/10.1007/978-3-030-13342-9_21

important sites are Bitcoin Analytics, CoinDesk, Cryptohopper and AvaTrade[1]. In this paper, we investigate whether strong or weak form of arbitrage strategies are indeed possible by trading across different Bitcoin Exchanges. In Sect. 2 we describe the dynamics of Bitcoin price by applying the well-known Black and Scholes model (see [5]) in a multi-variate fashion; under this assumption a strong arbitrage opportunity exists in the market even if the strategy is performed with discrete time revision of the portfolio. This theoretical arbitrage might be much outperformed in practice; in Sect. 3 we present an example of arbitrage strategy which take also advantage of the bid-ask spread mismatch in market Exchanges.

2   Modeling the Bitcoin Price Dynamics and Arbitrage Opportunities

2.1   Data Description

Bitcoin is traded in multiple online trading platforms (Exchanges), where different exchange rates are applied against the same fiat currency. Even ignoring market frictions such as the bid/ask spread, by considering the mid-price as the unique price, still there are multiple prices for Bitcoin available from different Exchanges. We consider daily prices from 01/01/2015 to 31/12/2017 available on the Exchanges website for Bitstamp, Gdax, Kraken, CEX.IO and BitKonan[2] and the corresponding value of the BlockChainInfo Index[3]. In Fig. 1 we plot the above Bitcoin prices and Index in US Dollar (USD) over the whole time series and two selected sub-periods to better appreciate the different exchange rates.

As we can notice from Fig. 1, the behavior of Bitcoin price in the different Exchanges is pretty the same. Indeed, the correlation between time series of Bitcoin prices obtained in the five considered Exchanges is approximately perfect, since it is close to 1.

In this paper, we assume that there are I Exchanges trading Bitcoin in the same fiat currency (i.e. USD) and denote by $S_t^{(i)}$ the price of one Bitcoin quoted in Exchange i at time t. In order to take into account the almost perfect correlation between the different Exchanges, we consider a common risk-source for all platforms. More precisely, we assume that the price dynamics of Bitcoin is described, in every Exchange, by the well-known Black and Scholes model [5], where

$$dS_t^{(i)} = \mu_i S_t^{(i)} dt + \sigma_i S_t^{(i)} dW_t, \quad S_0^{(i)} = s_0^{(i)} > 0. \tag{1}$$

Here, for every $i = 1, 2, \ldots, I$, μ_i, $\sigma_i > 0$, represent model parameters, $W = \{W_t, \ t \in [0, T]\}$, is a standard Brownian motion modeling the fluctuation of

---

[1] Available at: http://bitcoin-analytics.com/, https://www.coindesk.com/price/, https://www.cryptohopper.com/, https://www.avatrade.com/forex/cryptocurrencies/bitcoin.

[2] Available at: https://www.bitstamp.net/, https://www.gdax.com/, https://www.kraken.com/, https://cex.io/, https://bitkonan.com/.

[3] http://www.blockchain.info/.

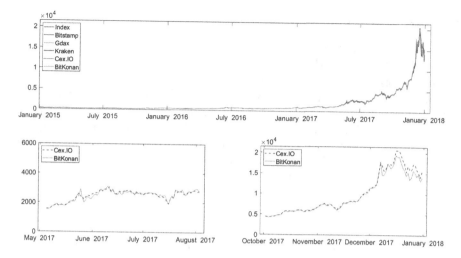

Fig. 1. The Bitcoin price in USD according to 5 different Exchanges and the Index value (top) and two sub-samples with only two exchange rates (bottom).

the Bitcoin market and $T > 0$ denotes a fixed finite time horizon. Note that the process $\mu_i S_t^{(i)}$ is the so-called drift coefficient, which represents the "average appreciation" of the process $S^{(i)}$ at time t, while $\sigma_i S_t^{(i)}$ is the diffusion coefficient, measuring the intensity of the source of randomness given by W.

Note that, the Bitcoin price processes $S^{(1)}, \ldots, S^{(I)}$ are perfectly correlated since they have the same "driving" Brownian motion W; in addition, different Exchanges are characterized by (possibly) different parameters values in the dynamics.

In Fig. 2 we plot two possible paths for two months of daily prices simulated according to model in (1) with parameters $\mu_1 = \mu_2 = 1.5$ and $\sigma_1 = 0.75, \sigma_2 = 0.9$. The picture exhibits a similar pattern to the one observed in Fig. 1.

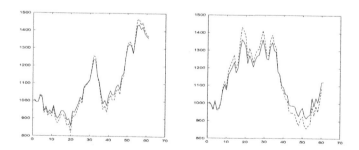

Fig. 2. An example with two simulated paths for two months of daily prices: parameters are set to $\mu_1 = \mu_2 = 1.5$ and $\sigma_1 = 0.75$ (solid), $\sigma_2 = 0.9$ (dashed) respectively.

2.2 Arbitrage Opportunities

To show that arbitrage is theoretically possible, we focus on two different Exchanges and denote the corresponding Bitcoin prices as $S^{(1)}$ and $S^{(2)}$. In our framework, we have $S_t^{(i)}$ with $i = 1, 2$.

Let us denote by r the constant risk-free rate. The risk premia for the the Bitcoins quoted in Exchanges $i = 1, 2$ is defined as $\text{RiskPremium}_i = \mu_i - r$.

The risk premium can be seen as the return in excess of the risk-free rate of return an investment in Bitcoin is expected to yield. Then, the corresponding *Sharpe ratios*, that is, the average returns earned in excess of the risk-free rate per unit of volatility or total risk, are given by

$$\text{SharpeRatio}_i = \frac{\text{RiskPremium}_i}{\sigma_i} = \frac{\mu_i - r}{\sigma_i}, \quad i = 1, 2.$$

Note that two assets that are perfectly correlated as $S^{(1)}$ and $S^{(2)}$, must have the same Sharpe ratio. Otherwise, it is possible to exploit an arbitrage opportunity, i.e. a trading strategy with zero initial outlay and a non-negative future payoff (so it does not expose to any risk) that is positive with positive probability.

This means that a sufficient condition to realize this investment strategy is $\text{SharpeRatio}_1 > \text{SharpeRatio}_2$.

Indeed, let us consider the self-financing portfolio $(\alpha^1, \alpha^2, \beta)$, where, for any $t \in [0, T]$, we buy the amount $\alpha_t^1 = C \left(S_t^{(1)} \sigma_1 \right)^{-1}$ of Bitcoin with price $S^{(1)}$ on Exchange 1, we short-sell the quantity $\alpha_t^2 = C \left(S_t^{(2)} \sigma_2 \right)^{-1}$ of Bitcoin with price $S^{(2)}$ on Exchange 2 and we invest/borrow the risk-free bond in the amount of the cost difference $C \left(\dfrac{1}{\sigma_1} - \dfrac{1}{\sigma_2} \right)$, where C is an arbitrary positive constant.

If V_t denotes the corresponding portfolio value at time t, then $(\alpha^1, \alpha^2, \beta)$ is a strategy with null initial value, since

$$V_0 = -\alpha_0^1 S_0^{(1)} + \alpha_0^2 S_0^{(2)} + \beta_0 = C \left(-\frac{1}{\sigma_1} + \frac{1}{\sigma_2} + \frac{1}{\sigma_1} - \frac{1}{\sigma_2} \right) = 0.$$

Moreover, the return of the above strategy is

$$dV_t = \alpha_t^1 dS_t^{(1)} + \alpha_t^2 dS_t^{(2)} - r\beta_t e^{rt} dt$$

$$= \frac{C}{S_t^{(1)} \sigma_1} dS_t^{(1)} - \frac{C}{S_t^{(2)} \sigma_2} dS_t^{(2)} - rC \left(\frac{1}{\sigma_1} - \frac{1}{\sigma_2} \right) dt$$

$$= C \left(\frac{\mu_1 - r}{\sigma_1} - \frac{\mu_2 - r}{\sigma_2} \right) dt = C \left(\text{SharpeRatio}_1 - \text{SharpeRatio}_2 \right) dt > 0,$$

and therefore it gives rise to an arbitrage opportunity since it produces a certain profit that is strictly greater than 0. The total gain of the strategy in the time interval $[0, s]$, for $s > 0$, is given by $C \left(\frac{\mu_1 - r}{\sigma_1} - \frac{\mu_2 - r}{\sigma_2} \right) s$. Here, C represents a scale factor which may leverage the total gain. Note that the above arbitrage opportunity is due to perfect correlation of the two dynamics. This is not the case in traditional financial markets; though some common risk factors may be identified to describe the systematic fraction of the variance of each asset, the idiosyncratic part of the variance is non negligible.

Example. Assume that parameters are set to $\mu_1 = \mu_2 = 1.5$, $\sigma_1 = 0.75$, $\sigma_2 = 0.9$ and $r = 5\%$ and that at time $t = 0$ we have $S_0^{(1)} = 1000$, $S_0^{(2)} = 1005$ and $C = 1000$ USD. Then, the arbitrage strategy consists in buying $\alpha_0^1 = 1.3086$ Bitcoins in Exchange 1, selling $\alpha_0^2 = 1.1056$ in Exchange 2 and an investment of 222.22 USD in the money market account. The initial cost of the investment is exactly V_0 by construction, while at time $t = \frac{1}{365}$ (one day) the profit is $\frac{1000}{365}$ (SharpeRatio$_1$ − SharpeRatio$_2$) = 0.88 USD. This is an arbitrage. Investments quotes should be revised in continuous time to keep the profit riskless.

2.3   Experiments

In what follows we consider the time interval from 01/01/2015 to 31/12/2017 as a training time for the model and we estimate the above model on daily Bitcoin prices available on Gdax, Bitstamp, Kraken, CEX.IO and BitKonan.

Table 1. Parameters fit of Black and Scholes model.

	Bitstamp	Gdax	Kraken	Cex.IO	BitKonan
μ_i	1.5313	1.6363	1.5317	1.5404	1.6770
σ_i	0.7323	0.8947	0.7451	0.7065	0.9139
SharpeRatio$_i$	2.0911	1.8289	2.0557	2.1803	1.8350

In Table 1 we report parameter values estimated on the above Exchanges. For the sake of simplicity, we have set $r = 0$ in the Sharpe ratios row.

First, we observe that parameter estimates are different across Exchanges, meaning that arbitrage opportunities arise; Gdax and CEX.IO Exchanges exhibit the most diverging Sharpe ratios.

The theoretical trading strategy considered in Sect. 2.2 is tested over the next 90 days by computing the overall daily profit of the investment applied to Gdax and CEX.IO; it is worth noticing that the strategy is performed by applying daily revisions of the investment rather than in continuous time as suggested in the theoretical model. Nevertheless it is still very profitable.

As an example, we plot in Fig. 3 the total theoretical profit obtained by investing according to the above strategy on Gdax and CEX.IO Exchanges which exhibit the largest difference in the Sharpe Ratio values. The Overall Gain is computed from January, 1 to march, 30, 2018 for $C = 100$. Clearly, by choosing a higher scale C the total gain increases proportionally.

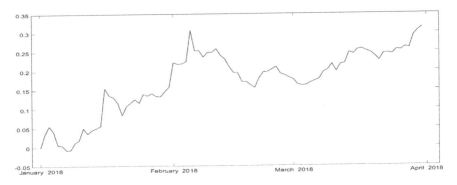

Fig. 3. Total profit dynamics in USD from January, 1, 2018 to March 30, 2018 (90 days), investing on Gdax and CEX.IO Exchanges with an initial investment $C = 1\$$.

3   Arbitrage Opportunities in Practice

To understand if there are arbitrage opportunities in practice we decided to make an application. Our application (TradeBitcoin), part of the suite BlockChain-Vis[4] [2–4] used for Bitcoin analysis and visualization, is based on finding the price options on the Bitcoin exchange and writing possible arbitrage operations on a database to see if it is possible to correctly perform an arbitrage on the Bitcoin market. To perform this, we choose the principal Bitcoin Exchanges that allow the use of web API to get Bitcoin prices in real time. In our application we used: Bitstamp, Kraken, CoinBase, ANX, Bitbay, Bitfinex, BitKonan, HitBTC and TheRock. In Fig. 4 we can see a table (see footnote 4), from our application, that shows in real time the price of all considered exchanges. Instead in Fig. 5 a chart (see footnote 4), from our application, that shows the number of times the BID has been greater and the ASK lower on a exchange in the last year.

Site	Bid	Ask	Fee(ask/bid)
Bitstamp	7640.12 $	7643.99 $	0.25 %
Kraken	7647.2 $	7652.9 $	0.34 %
CoinBase	7645.8 $	7667.45 $	1 %
ANX	7576.52387 $	7588.85163 $	0 %
Bitbay	7104.28 $	7500 $	0.43 %
Bitfinex	7556.2 $	7557 $	0.2 %
BitKonan	7511.31 $	7603.05 $	0.29 %
HitBTC	7581.82 $	7584.27 $	0.1 %
TheRock	7201.53 $	7543.67	0.2 %

Fig. 4. Exchange prices table on June 3, 2018.

---

[4] http://normandy.dmi.unipg.it/blockchainvis/.

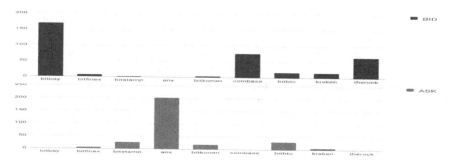

Fig. 5. Number of times that BID has been greater and ASK lower last year.

3.1 How It Works

Starting with an initial endowment of 1000 USD in November 2015, we build an investigation strategy that:

– Obtains with a fixed time frequency, bid and ask prices by the aforesaid exchanges.
– Checks and compares BID and ASK values in different exchanges, also controlling for the **amounts offered** and **exchange's fees**.
– has a frequency of 6 h.
– spends maximum of 100 USD per transaction.
– spends maximum of 100 USD per day.
– makes transactions only with a minimum gain of 2%
– simulates the arbitrage position and the corresponding gain added to our portfolio value.

3.2 Results

The evolution of the portfolio value (see footnote 4) (Fig. 6) lead to a total outcome of about 3500 USD in October 2017 that is a 250% return in less than

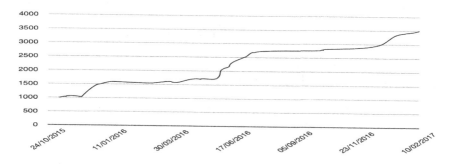

Fig. 6. Our gain (in USD) over time.

two years with no risk! By relaxing some of our constraints we could obtain a still better performance. Of course, there are practical issues in order to make this arbitrage a real strategy; instead for many exchanges, it is not possible to do real-time payments from a selected bank account so we need to store a certain amount of USD in all the exchanges we are wishing to invest on. Further the Bitcoin trading is not instantaneous, but need many minutes to be completed. Now, it is also possible to make immediate money transfers[5] with an average cost of to 2 USD, so if we subtract these costs to our profit we still have 2000 USD.

4   Conclusion and Future Work

Bitcoin is a cryptocurrency that does not require a third-party intermediary neither a central authority. Therefore, the Bitcoin network is decentralized, and transactions are performed by the users of the system i.e. peer to peer. Bitcoin volatility is high, hence it has been claimed as a speculative financial asset rather than a currency. In addition, arbitrage opportunities are indeed possible by trading on different Exchanges, as we prove both via a theoretical model and via an empirical strategy. Our approach is complementary to other theoretical studies on Bitcoin arbitrage such as [1,9] where the authors study triangular arbitrage with Bitcoin, i.e. buying bitcoin in US dollars and selling them in Chinese Renminbi. Further attention will be devoted in our future research to possible theoretical arbitrages by considering the model introduced in [7] as a starting point for a multi-exchange approach.

References

1. Barker, J.A.: Triangular arbitrage with Bitcoin. Ph.D. thesis (2017)
2. Bistarelli, S., Mercanti, I., Santini, F.: An analysis of non-standard bitcoin transactions. In: Crypto Valley Conference on Blockchain Technology. IEEE Computer Society (2018, to appear)
3. Bistarelli, S., Parroccini, M., Santini, F.: Visualizing bitcoin flows of ransomware: wannacry one week later. In: Proceedings of the Second Italian Conference on Cyber Security (ItaSEC). CEUR Workshop Proceedings, vol. 2058. CEUR-WS.org (2018)
4. Bistarelli, S., Santini, F.: Go with the -bitcoin- flow, with visual analytics. In: Proceedings of the 12th International Conference on Availability, Reliability and Security (ARES), pp. 38:1–38:6. ACM (2017)
5. Black, F., Scholes, M.: The pricing of options and corporate liabilities. J. Polit. Econ. **81**, 637–654 (1973)
6. Böhme, R., Christin, N., Edelman, B., Moore, T.: Bitcoin: economics, technology, and governance. J. Econ. Perspect. **29**(2), 213–238 (2015)

---

[5] https://www.europeanpaymentscouncil.eu/what-we-do/sepa-instant-credit-transfer.

7. Cretarola, A., Figà-Talamanca, G., Patacca, M.: Market attention and Bitcoin price modeling: theory, estimation and option pricing (2017). Available at SSRN: https://papers.ssrn.com/sol3/papers.cfm?abstract_id=3042029
8. Nakamoto, S.: Bitcoin: a peer-to-peer electronic cash system. http://www.hashcash.org/papers/hashcash.pdf (2008). Accessed 28 Jan 2018
9. Pieters, G., Vivanco, S.: Bitcoin arbitrage and unofficial exchange rates (2015)
10. Yermack, D.: Is bitcoin a real currency? An economic appraisal. In: Handbook of Digital Currency, Chap. 2, pp. 31–43. Elsevier (2015)

Author Index

Printed in the United States
By Bookmasters